Economy and Social Inclusion

Creating a Society for All

Series Editor

Akihiko Matsui, Tokyo, Japan

This series provides a forum for theoretical, empirical, historical, and experimental analysis of issues related to economy and social inclusion and exclusion. Included are the interconnected problems of alienation, deprivation, discrimination, economic inequality, polarization, and prejudices caused by or related to abusive behavior, aging, depopulation, disability, the educational gap, the gender gap, natural disaster, poverty, rare diseases, war, and various other economic and social factors.

The common theme of the series is to examine how we can create an inclusive society that accommodates as many people as possible and promotes their welfare. We believe this should be the main goal of economics as a discipline. Society need not passively observe its inequality, social exclusion, and decline. Policy, institutions, and our actions matter. The series aims to enrich academic discourse, influence economic and social policy, and enlighten a global readership.

Methodologies adopted in this series are mainly economic theory, game theory, econometrics, statistical analysis, economic experiments, and disability studies. Psychology, sociology, legal studies, and medical sciences, among other disciplines, are also considered important related fields of study.

The interdisciplinary research taken up in the series utilizes these existing methodologies for the common goal of creating a society for all. Furthermore, progress in such interdisciplinary studies will contribute new insights to the development of economic studies. The major geographical targets of the series are Japan and other Asian countries but are not restricted to those areas. At the same time, however, the goal is to amplify the findings therein to universally applicable insights and knowledge.

Series Editor:

Akihiko Matsui (Professor, The University of Tokyo, Japan)

Editorial Board:

William P. Alford (Vice Dean for the Graduate Program and International Legal Studies, Jerome A. and Joan L. Cohen Professor of East Asian Legal Studies, Harvard Law School)
In-Koo Cho (Professor, University of Illinois, USA)
Partha Sarathi Dasgupta (Frank Ramsey Professor Emeritus of Economics at the University of Cambridge, UK)
Hidehiko Ichimura (Professor, The University of Tokyo, Japan)
Daiji Kawaguchi (Professor, The University of Tokyo, Japan)
Osamu Nagase (Visiting Professor, Ritsumeikan Univesity, Japan)
Yasuyuki Sawada (Chief Economist, Asian Development Bank, Philippines; Professor, The University of Tokyo, Japan)
Tomomi Tanaka (Senior Economist, The World Bank)

More information about this series at http://www.springer.com/series/13899

Akihiko Matsui

Economy and Disability

A Game Theoretic Approach

Springer

Akihiko Matsui
Faculty of Economics
University of Tokyo
Tokyo, Japan

ISSN 2509-4270   ISSN 2509-4289   (electronic)
Economy and Social Inclusion
ISBN 978-981-13-7625-2   ISBN 978-981-13-7623-8   (eBook)
https://doi.org/10.1007/978-981-13-7623-8

This Springer imprint is published by the registered company Springer Nature Singapore Pte Ltd.
The registered company address is: 152 Beach Road, #21-01/04 Gateway East, Singapore 189721,
Singapore

Preface

In undertaking the work toward this book, I have been supported by many people, including some with disabilities. Unfortunately, I cannot list all of them. Part I of the book discusses various cases and I have met the people appearing in most cases, and collaborated with them in others. Some of them are mentioned here. Professor Shinichiro Kumagaya educated me on the relationship between dependence and independence in disability. Ms. Kumiko Hamazaki, the Director of Kanamachi-gakuen, introduced me to a setting where deaf people are the majority. The students of Taira School for handicapped children taught me the importance of supporting others while being supported. Those with different color perception taught me how society is designed for the "ordinary" people, while Ms. Fusae Kurihara taught me about the hardship she encountered as a nurse with a hearing disability. Ms. Noriko Segawa and Ms. Kumiko Usui taught me about disqualifying clauses on disability. Dr. Masahiro Kami educated me on rare diseases. Ms. Yurie Yoshino showed me how her cancer is a "forgotten cancer" because it is so rare when compared with other major cancers. Dr. Yuko Kodama, a nurse, explained to me the financial burden faced by patients with chronic myeloid leukemia. Mr. Sotaro Masunaga, the President of Masunaga Optical Mfg. Co., Ltd., taught me about his firm founded by his great grand father, who introduced the eyewear industry to Fukui. Mr. Hiroshi Imanaka, the Director of Atelier Incurve, showed me how he and the artists of the atelier had created an art market for themselves, while, in a more mainstream facility, Mr. Susumu Murakami, the owner/manager of a *juku*, showed me how a *juku* can be an asylum or refuge for children.

Part III of the book is based on my published works, and many of these works have been collaborations. I greatly appreciate the insight and assistance of my collaborators and thank them for allowing these works to be included in the present book. Professor Itzhak Gilboa was not only my collaborator, but also my thesis adviser at Northwestern University. Professor Masahiro Okuno-Fujiwara was my undergraduate adviser in Japan. Professor Mamoru Kaneko was a mentor in my research, and Dr. Kimberly Katz was a graduate student under my supervision at the University of Pennsylvania. Many thanks go to each of them.

I would like to thank other collaborators of mine, including Professors Nobuhiro Kiyotaki, Fumio Hayashi, Kazuhiko Kato, Toshiji Kawagoe, Roger Lagunoff, Yiting Li, Cesar Martinelli, Kiminori Matsuyama, Daisuke Oyama, Andrew Postlewaite, Takashi Shimizu, and Tetsuo Yamamori. Professor In-Koo Cho has been my collaborator for nearly 30 years, publishing more than 5 papers together.

This book adopts evolutionary game theory as an essential building block. In particular, Sect. 8.4 explains why we need to adopt deterministic dynamics rather than stochastic dynamics to understand disability-related issues. I thank Professors Drew Fudenberg, Joseph Hofbauer, Michihiro Kandori, George Mailath, Rafael Rob, and Bill Sandholm have inspired me on this subject.

I also would like to thank my mentors, Professors Motoshige Itoh, Ehud Kalai, and Ariel Rubinstein.

Much of the work in this book is based on research that was performed after the formation of a research team on disability and economy. The research team was made up of a loose association of about 30 researchers, many of whom have a disability (I recently also became a person with a disability). I am grateful to all of them, and especially to Profs. Osamu Nagase and Satoshi Fukushima, who introduced me to disability studies. Professor Satoshi Kawashima has worked for the team with his talent and enthusiasm. Professor Ryoji Hoshika taught me disability studies. Professor Mai Yamashita educated me about the history of disability in Japan. Our team conducted several surveys. I am grateful to the members of the survey team, especially Professors Ryoko Morozumi, the leader of the first survey, Dr. Yoshihiro Kaneko, the leader of the second survey, Drs. Kazuko Kano, Machiko Kawamura, Soya Mori, Akira Nagae, Emiko Tanaka. Ms. Chieko Doi, Ms. Hizuki Moriwaki, and Ms. Hiromi Tojima have supported me in organizing and maintaining the team on disability and economy. Mr. Taichi Niwa, Mr. Tomoya Ozeki, Mr. Yoshiki Tomita, and Mr. Takeshi Tsuchiya have supported me, working home as it is hard for them to commute. Special thanks are also given to Prof. Masahiko Aoki, who suggested that I should apply for a Grant-in-Aid from the Japan Society for the Promotion of Science (JSPS) to start this project. Without his suggestion and encouragement, the research team on disability and this book would never have existed. A sequence of Grants-in-Aid from JSPS is gratefully acknowledged.

Thanks are also due to those who directly supported me in completing this book. Ms. Megumi Murakami provided great assistance in completing the bibliography as well as editing the mathematical parts of the book, while the English text of the draft version was edited by Dr. Austin Schultz. Ms. Juno Kawakami of Springer Nature provided patient encouragement throughout this project from beginning to end. All remaining errors, of course, are mine.

Last but not least, I express my profound thanks and appreciation to my family members, who have supported me throughout the various stages of my life. Without their support, I could not have completed this work.

Tokyo, Japan   Akihiko Matsui

Contents

1   **Introduction**   1
     1.1   The Escalator in Japan   1
     1.2   Some Basic Facts About Disability in Japan   4
     1.3   A Game-Theoretic Analysis   8
     1.4   Organization of the Book   12

Part I   Disability and the Market: Cases

2   **The Market for All?**   17
     2.1   Independence of Persons with Disability   17
     2.2   Supporting Others While Being Supported   20

3   **Economies for the Majority**   23
     3.1   Children with Hearing Disability   23
     3.2   The Dancer Fighting with a "Forgotten Cancer"   25
     3.3   Chronic Myelogenous Leukaemia   27
     Appendix: Simple Simulation for the Welfare Loss Caused
     by Incomplete National Insurance System   29

4   **Dominant Views and Prejudices**   33
     4.1   "There Is No Such Color!"   33
     4.2   Disqualifying Clauses on Disability   35

5   **Creating the Market for All**   39
     5.1   The City of Eyeglasses   39
     5.2   Atelier Incurve   41
     5.3   Children Attending *Juku*   46

Part II   Game Theory and Disability

6   Game Theory   51
    6.1   Introduction   51
    6.2   Social Norms and Rational Behavior   52
    6.3   Games in Strategic Form   53
        6.3.1   Games in Strategic Form and Zero-Sum Games   53
        6.3.2   Nash Equilibrium   55
    6.4   Evolutionarily Stable Strategies   56
    6.5   Deterministic Dynamics   58

7   The More, the Better   61
    7.1   Scale Economy, Coordination Games, and Strategic
        Complementarity   61
    7.2   Network Externality   64
    7.3   Institutional Complementarity   65

Part III   Toward a Theory of Economy and Disability

8   Best Response Dynamics   69
    8.1   Introduction   69
    8.2   Social Stability   71
        8.2.1   Definitions and Notations   72
        8.2.2   Socially Stable Sets   73
        8.2.3   Properties of Socially Stable Sets   76
    8.3   Static and Dynamic Concepts of Social Stability   79
        8.3.1   Static versus Dynamic Concepts   79
        8.3.2   Definitions and Notations   81
        8.3.3   Evolutionarily Stable Strategies and Replicator
            Dynamics   82
        8.3.4   A Point-Valued Solution Concept and Its Static
            Equivalence   83
        8.3.5   Set-Valued Solution Concepts and Their Existence   86
        8.3.6   Various Games   88
    8.4   Deterministic versus Stochastic Dynamics   91

9   Cheap-Talk and Cooperation in a Society   95
    9.1   Introduction   95
    9.2   Games of Common Interest with Cheap-Talk   98
    9.3   Social Stability and Socially Stable Sets   100
    9.4   Optimality Result and Its Proof   101
    9.5   Other Games   104
    9.6   Voice Matters in a Dictator Game   106

10 Evolution and the Interaction of Conventions 107
  10.1 Introduction 107
  10.2 The Model 109
    10.2.1 No-coordination Equilibria 110
    10.2.2 Partial Coordination 112
  10.3 The Evolution of Conventions Through Integration 113
  10.4 Welfare 115
  10.5 Extensions 115
    10.5.1 Endogenous Matching Probability 115
    10.5.2 Government Intervention 117
    10.5.3 Discrimination 118
    10.5.4 A K-Society Model 119
  10.6 Remarks 120

11 When Trade Requires Coordination 123
  11.1 Introduction 123
  11.2 The Model 126
  11.3 Equilibria 128
    11.3.1 Autarky 128
    11.3.2 Unification 130
  11.4 Dynamics 132
    11.4.1 Small Community Versus Large Community 132
    11.4.2 Incomparable Community Sizes 135
    11.4.3 Different Initial Conditions 135
  11.5 Welfare Implications 136
  11.6 Welfare Under Increasing-Returns-to-Scale Matching
       Technology 138
  11.7 Remarks 142

12 A Model of Man as a Creator of the World 145
  12.1 Introduction 145
  12.2 Inductive Construction of Models 148
    12.2.1 Impressions and Experiences 148
    12.2.2 Models 148
    12.2.3 Axioms 148
    12.2.4 Prior Beliefs 150
  12.3 Applications 150
    12.3.1 Predation 150
    12.3.2 Bullying 152
    12.3.3 Pioneers 153
  12.4 Induction and the Science of Man 154
    12.4.1 Hume 154
    12.4.2 Einstein 156
  12.5 Remarks 157

13 Segregation and Discrimination in Festival Games 159
   13.1 Introduction 159
   13.2 Festival Games 163
   13.3 Experiences and Inductive Stability 165
   13.4 Segregation and Discrimination 169
       13.4.1 An Example 171
   13.5 Escaping from Segregation 174

14 Prejudices Induced by Segregation 179
   14.1 Introduction 179
   14.2 Inductive Construction of an Image of the Society 179
       14.2.1 Individual Models Built by a Player 180
       14.2.2 Coherence of Models with Experiences 182
   14.3 Rationalization 184
   14.4 Prejudicial Models 184
   14.5 Escaping from Prejudicial Models 188
   14.6 Discussion 189
       14.6.1 Sequential Rationality and Rationalization 190
       14.6.2 Interactions Between Models and Behavior 191
       14.6.3 Comparisons with Merton's Classification 193
       14.6.4 Large Societal Games Versus Small Micro Games 194
   Appendix 195

15 Everyone on the Island Spoke Sign Language 199
   15.1 Introduction 199
   15.2 Hereditary Deafness on Martha's Vineyard Island 202
   15.3 Model 203
       15.3.1 The Majority Bargaining Games 203
       15.3.2 The Unanimity Bargaining Games 204
   15.4 Analysis of the Second Stage: A Deductive Approach 206
       15.4.1 The Majority Bargaining Games 206
       15.4.2 The Unanimity Bargaining Games 210
   15.5 The Analysis of the First Stage: An Evolutionary Approach . . . 211
       15.5.1 The Majority Bargaining Game 212
       15.5.2 The Unanimity Bargaining Game 213
   15.6 Prejudice: An Inductive Approach 214
   15.7 Remarks 217

16 Concluding Remarks 219

References 223

Author Index 229

Subject Index 233

About the Author

Akihiko Matsui is Professor of Economics at the University of Tokyo. He is a fellow and a former council member of the Econometric Society. He also served as the President of the Japanese Economic Association (JEA) in 2016–2017. Professor Matsui obtained his Ph.D. at Northwestern University and held positions at the University of Pennsylvania and the University of Tsukuba. His fields of study include game theory, monetary economics, and studies on economy and disability. Professor Matsui has published papers in *Econometrica*, *Economic Theory*, *Experimental Economics*, *Games and Economic Behavior*, *International Economic Review*, *International Journal of Industrial Organization*, *Japanese Economic Review*, *Journal of Economic Theory*, *Journal of the Japanese and International Economies*, *Journal of Public Economic Theory*, *Mathematical Social Sciences*, *Review of Economic Design*, *The Review of Economic Studies*, and others. He is a previous recipient of the *Nikkei* Book Award, the Nakahara Prize of JEA, the JSPS Prize, and the Japan Academy Medal.

Chapter 1
Introduction

The purpose of this book is to present a game-theoretic approach to issues related to economy and disability. This introduction consists of four sections. The first section presents relevant anecdotes to give the reader some context, while the second section presents some basic facts on disability in Japan. Although we do not intend to confine our attention to the disability-related issues of Japan, we need to share very basic facts on disability and take Japan as an example particularly because cases in this book take place in Japan. A cross-country description can be found in OECD (2003). The third section briefly describes game theory and its use and discusses the importance of formal models in studying disability-related issues. The fourth section explains the organization of this book.

1.1 The Escalator in Japan

The use of escalators in Japan is fascinating to watch. This interest does not relate to the physical aspects of the escalators, but to the way people ride on them. Typically, some people stand still on the escalator, while others choose to walk forward. Japanese people have established a convention whereby those who stand still do so on one side of the escalator, while those who walk forward use the other side. Different regions have established different conventions. In Tokyo, people stand on the left side and walk forward on the right. In Osaka, however, the second largest city in Japan, people stand on the right side and walk forward on the left.

Given the Japanese moral that you should not stand in someone's way, both Tokyo and Osaka styles are "stable" conventions. If everyone stands on the left (resp. right) side, you have an incentive to stand on the left (resp. right). Essentially, it does not matter if people stand on the left side or the right side; what matters is that you stand on the same side of the escalator as the other users. Once such a convention is

© Springer Nature Singapore Pte Ltd. 2019
A. Matsui, *Economy and Disability*, Economy and Social Inclusion,
https://doi.org/10.1007/978-981-13-7623-8_1

established, it continues for a long time unless there is a disturbance from outside. People follow this convention without thinking much about it.[1]

Recently, a significant disturbance to this convention has emerged. Local authorities decided to impose a new rule that required all users to stand still on both sides of the escalator, and people should not walk or run. This rule was a response to some accidents that occurred between users standing on the escalator and others walking past. This move was an attempt to change a convention that has been spontaneously established for a long time. One may call it a battle between the convention and the rule.

People begin to think about the convention that they have followed unconsciously once a new rule is set against it. Even after the new escalator rule was announced, it is not important when the escalator is sparsely used. Many people in Tokyo continue to stand on the left, and some people occasionally walk on the right. No conflict occurs between the convention and the new rule when the escalator is overly crowded, because people stand side by side so that there is no room for others to walk through. A problem arises when the escalator is slightly crowded. It is more efficient that the users stand side by side, but many people are reluctant to be the first person to stand on the right (in Tokyo) with the concern that someone will approach from behind. The expectation, however, is that once someone dares to stand on the right, others will follow suit and stand on the right to be followed by other users, and so forth.

We are surrounded by various, innumerable conventions, including morning greetings, table manners, and proper attire. Conventions economize our thinking—it is easy to follow conventions rather than considering everything from the beginning. We sometimes get tired in new environments because we have to learn the way people behave, which takes extra time and effort.

One of the main properties of conventions is that the more people follow a convention, the likelier it is that you are better off following it as well. You should bow if everyone bows, or you should shake hands if everyone shakes hands.

However, people do not blindly follow conventions. It may be true that people are not conscious of many conventions that are used in our daily lives, but, when they are asked to follow a particular convention, people will make decisions by taking various things into account. If a convention is shaken, people will consider whether they should take a new action or not.

A convention does not necessarily change even if the corresponding rule changes. In particular, a convention tends to be maintained if people's minds remain unchanged. In the 2010s, various Japanese laws on disability-related issues were amended and enacted, including the Act on Employment Promotion etc. of Persons with Disabilities, Basic Act for Persons with Disabilities, and Act for Eliminating Discrimination against Persons with Disabilities. This move was in response to the United Nations' Convention on the Rights of Persons with Disabilities (CRPD), which was ratified in 2014. For example, in the Act for Eliminating Discrimination against Persons with Disabilities, the term "reasonable accommodation," one of the

---

[1]The only exception may be Shin-Osaka station where there are lots of people coming from Tokyo so that people get confused whether they should stand on the left or right side of the escalator.

key concepts in CRPD, was introduced in addition to direct and indirect discrimination. In spite of these laws, discrimination continues even in the government. For example, according to *Asahi Shimbun*, one of the leading Japanese newspapers, at least 28 prefectural governments and several central government agencies have inserted the following clause (or similar) in their job application guidelines for persons with disabilities: "the applicant has to be able to commute by him/herself and to work without assistance," (Oct. 27, 2018).

Satoshi Kawashima at Okayama Science University, who specializes in the rights of persons with disabilities, says:

We should regard such conduct 'unconscious discrimination' even if there is no intention of discriminating against them. The Act on Employment Promotion etc. of Persons with Disabilities was amended several years ago, and this incident is an evidence that prejudice and ignorance have remained in our society. The change of laws is meaningless unless it is followed by the real change (*Kyoto Shimbun*, Nov. 2, 2018).

The central and local governments are likely to modify the job application clause in response to these news reports. However, it is possible that they reject persons who cannot commute by themselves unless they have a deliberate intention of hiring them. If that is the case, the change in rules would end up being tokenistic.

Another striking news report emerged in the summer of 2018. According to the Japan Times, one of the leading English newspapers in Japan:

Nearly ten ministries are suspected of having inflated their employment rates (of people with disabilities) routinely for over 40 years. The revelation is likely to spark criticism of the government given that it has long called on the private sector to hire more people with disabilities (*Japan Times, Aug. 17, 2018*).

The Japanese government has introduced the employment quotas for persons with disabilities, and they are imposed upon the government as well as the private sector. From the viewpoint of economics, this quota system is similar to the tax-subsidy system for the private sector. The quota is set at 2.2 % for the private sector and 2.5 % for the public sector. For the private sector, a company has to pay a fee of 50,000 yen per month for each person below the quota and receives a subsidy of 27,000 yen per month for each person above the quota. For example, if a firm employs 1,000 people, its quota becomes 22 employees with disabilities. If the firm employs 19 people with disabilities, then it must pay 150,000 yen (=50,000 yen × 3) per month. However, if the firm employs 25 people with disabilities, then it will receive a subsidy of 81,000 yen (27,000 yen × 3) per month. The public sector is not subject to this tax-subsidy system even though each government agency has to meet the quota. *Nikkei Shimbun*, the Japanese counterpart of the *Wall Street Journal* or *Financial Times*, reported that the actual number of the employment of persons with disabilities is 3407, which is less than half of the reported number of 6867 (Aug. 29, 2018). Moreover, this misreporting is widespread and systematic as shown in Table 1.1.

The central and local governments seem to have not taken the change of rules seriously, even though they are required to follow the law. Thus, they appear to have officially adopted the new rule at the surface level, while retaining the old convention

Table 1.1 Employment of persons with disabilities by major Japanese government agencies

Ministry/Agency	Actual number	Reported number
Foreign	25	150
Environment	15	46
Education	16	51
Revenue	389	1411.5
Infrastructure	286.5	890
Internal Affairs	40	110
Finance	94.5	264.5
Justice	262.5	802
METI	52	153.5
Defense	201	516
Cabinet Office	29	56
Agriculture	195.5	364
Meteorological	65	112
Total	3407.5	6867.5

Data taken from *Nikkei Shimbun* (Aug. 29, 2018). Numbers are not necessarily integers because of a specific way of counting. The names of agencies may be abbreviated.
METI Ministry of Economy, Trade, and Industry

at the practical level. The behavior of these agencies should be monitored to ensure that they do not to distinguish between law and practice.

New conventions are born with new minds and new behavior of people. The reform of conventions and minds is indispensable not only in daily problems such as escalator use as described above, but also in socio-economic problems such as employment promotion of persons with disabilities. The more users stand on both sides of the escalator, the likelier it is that the new convention will be established. Similarly, as more persons with disabilities work in an office, any discrimination against them will decline. Disability-related issues should be understood as a societal problem.

The anecdotes presented in this section motivate a game-theoretic approach. However, before moving onto game theory, we examine some basic descriptive statistics about persons with disabilities in Japan.

1.2   Some Basic Facts About Disability in Japan

Japan uses a disability registration system where persons with a disability must register at the respective government offices to be acknowledged as such. There are three basic disability categories: physical, intellectual, and mental.

According to the Annual Report on Government Measures for Persons with Disabilities (*Shogaisha Hakusho*) 2017 (Cabinet Office of Japan 2017b), 8,587,000 people (6.8% of the total population) are registered as persons with disability. Of

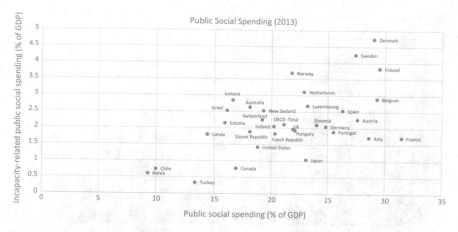

Fig. 1.1 Social expenditures (figure created by author based on OECD, 2013)

these, the numbers of registered persons with physical, intellectual, and mental disabilities are 3,922,000 (3.1%), 741,000 (0.6%), and 3,924,000 (3.1%), respectively.

Next, let us look at the government expenditures on disability-related policies. Figure 1.1 shows the social expenditures, the total and the disability-related expenditures as a percentage of gross domestic product (GDP) for OECD countries (OECD 2013). While Japan exceeds the average in terms of the total social expenditures, it falls far behind in terms of the disability-related expenditures.

Persons with disability are, *albeit* not without dispute, relatively well cared for when they are at school and/or at welfare institutions. For example, in all the elementary and high schools, there is one teacher for every 14.8 students, while there is one teacher for every 1.7 students in special education schools[2] However, once people with disabilities try to enter the economic arena such as the labor market, they often face prohibitive barriers without much public support.

Such a lack of public support for job participation is not difficult to observe. According to government documents concerning the budget of the policies for persons with disabilities in fiscal year 2017,[3] the total budget was about 1.93 trillion yen. Among this budget, support for welfare services was 1.32 trillion yen, health and medical support received 0.38 trillion yen, and job support, including support for financially independent living, was a mere 0.21 trillion yen.

To see further alarming data, let us look at budgetary data of disability-related expenditures by the Ministry of Health, Labor and Welfare.[4] The total budget of the Department of Health and Welfare for Persons with Disabilities in fiscal year 2017 was about 1.73 trillion yen, and the amount of about 1.27 trillion yen was allocated to expenditure on disability-related services. Among these figures, a total of 1.22

---

[2]Fiscal year 2014. See Ministry of Education Culture Sports Science and Technology (2015).
[3]See Cabinet Office of Japan (2017a).
[4]See Ministry of Health, Labour and Welfare (2017).

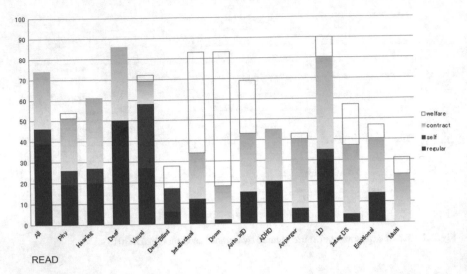

READ

Fig. 1.2 Labor force participation rates across disability categories (Survey by READ 2009–10)

trillion yen was used to provide persons with disabilities with the necessary services for living. Only 1.1 billion yen (0.0011 trillion yen) was allocated to the promotion of job participation for persons with disabilities.

Figure 1.2 shows job participation rates for ages 18–64 years across various categories of disability. It is based on a survey by our team, Research on Economy And Disability (READ) in 2009–10.[5]

To interpret Fig. 1.2, note first that in Japan, regular jobs and self-employed positions are often regarded as "decent jobs," especially for so-called bread-winners or primary providers. Contract-based jobs are often considered as secondary or supportive roles in terms of household income, while welfare jobs are for persons with disabilities. These welfare jobs are not subject to the minimum wage law so that their hourly wage rates can be extremely low, as the subsequent data shows.

The overall labor participation rates (excluding welfare jobs) of persons with disabilities are low compared to those of persons without disabilities.[6] About 45% of the population, including persons with or without disability, are regular workers or self-employed, with a labor force participation of about 75%. For those with physical

---

[5]READ had been supported by grants-in-aid (principal investigator: Akihiko Matsui). This survey (Matsui et al. 2012) was conducted from July 2009 to December 2010, using conventional mail. Samples were taken from the members of disability associations over age 18. The total of 2272 questionnaires were distributed, and 1331 were recovered (recover rate: 58%). The data for the entire population ("All" in the figure) is taken from *Shugyo-kozo-kihon-chosa* (Employment Status Survey) of Ministry of Internal Affairs and Communications.

[6]One exception is the category "Deaf." This is mainly caused by the sample bias where many samples for the deaf people are taken from the executive members of the prefectural branches of the deaf association.

disabilities, excluding hearing and visual ones, only about 25% hold a regular job or are self-employed, and 50% in total.

In addition, among persons with disabilities, there are clear differences between different categories. Physical disabilities, including hearing and visual disabilities, fare better than intellectual and mental disabilities. Let us look at intellectual disability. Table 1.2 shows the distributions of monthly salaries of persons with intellectual disabilities, and Fig. 1.3 uses data taken from this table for the total numbers. To compare, the minimum wage rate in Japan is about 700–800 yen per hour, which equates to a fulltime monthly salary (40 h, 4 weeks) of 112–128 thousand yen. The fraction of persons with an intellectual disability who earn more than 130 thousand yen is 5.1%. A striking fact is that 48.2% of the "workers" with intellectual disability earn less than ten thousand yen per month. Also, at *sagyousho*, or sheltered workshops, this number jumps up to 72.9%.

Table 1.2 Monthly salary of persons with intellectual disabilities

	0	0–10	10–30	30–50	50–70	70–100	100–130	130–150	150–	Unknown
Total	3.5	44.7	13.0	6.2	6.7	9.0	5.8	3.0	2.1	6.3
Severest	5.1	78.0	6.8	0.0	0.0	1.7	1.7	0.0	0.0	6.8
Severe	1.6	62.6	13.2	3.3	3.8	2.2	2.7	0.0	0.5	9.9
Medium	3.8	40.7	17.8	9.7	8.5	8.5	3.0	2.1	0.8	5.1
Regular/ contract	0.4	2.1	5.0	13.0	18.5	26.1	18.1	8.4	5.9	2.5
Welfare Work	2.2	70.7	16.5	1.8	0.9	0.2	0.0	0.0	0.0	7.7

Column headings show income categories in thousands of yen. Table data are percentages. Data from Ministry of Health, Labour and Welfare, 2005

Fig. 1.3 Monthly salary of persons with intellectual disabilities (percentage in each category, in thousand yen, Ministry of Health, Labour and Welfare)

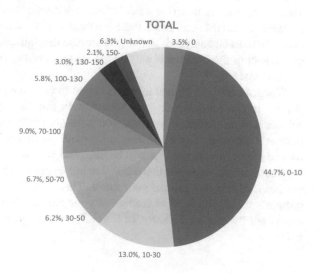

1.3   A Game-Theoretic Analysis

Although disability studies researchers have been concerned with economic issues, their notion of economics has often been confined to either institutional analysis in the case of institutional economics, or cost-benefit analysis in the case of neoclassical economics. Many non-economists who are interested in economic issues are often ignorant of the fact that economic analysis has been changed drastically in the past several decades, thanks to a new discipline called *game theory*, which we now turn to.

Game theory is a discipline that studies interactions between agents in a logically rigorous manner. "Agents" can be identified with many things. The most natural example of an agent is a human being. In economics, however, not only people but also firms are modeled as agents. In political sciences, nations can be identified with agents, and so on.

Game theory can consider a market system not as given but as an evolving institution. Using game theory, we argue that persons with disabilities suffer not necessarily because they are inferior, but because they are different from the majority of people.

One may think that human interaction is so complicated that it would be impossible that we study it in a coherent and logical manner. However, there are many scientific disciplines that study complicated phenomena. For example, it is hard to forecast the weather three months ahead. Similarly, it is difficult to predict human behavior in the near future. In this sense, natural sciences and social sciences are comparable in terms of difficulty of prediction. However, there is one huge difference between a weather forecast and the prediction of human behavior. This difference makes the latter more difficult to study than the former, but, at the same time, it makes game theory unique and fascinating.

In the case of a weather forecast, a subject and an object are dichotomized. On one hand, there is a scientist, or a weather forecaster, who would like to better understand the weather. On the other hand, there is a natural phenomenon called the weather that is out there to be analyzed. In many scientific disciplines, meteorology in this case, the subject or the scientist is outside the system he/she tries to analyze, understand, and forecast (Fig. 1.4).

When we analyze human interaction, we may face a problem. Suppose now that you try to predict your partner's behavior. In this case, you cannot dichotomize a subject and an object. If you do, you perhaps make mistakes in predicting your partner's behavior. You try to read what your partner thinks like the scientist who tries to read what the weather will be like, and you surely make mistakes. What you have to take into account is that while you are trying to read your partner's mind, your partner is also trying to read your mind. Therefore, in order to correctly read your partner's mind, you have to read what your partner thinks about you. Once again, this is not the end of the story. In Fig. 1.5, the two arrows start forming an infinite cycle:

Fig. 1.4 You think about the weather

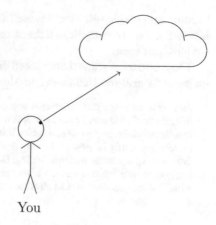

You think about the weather.

Fig. 1.5 You think about what your partner thinks about what you think about your partner

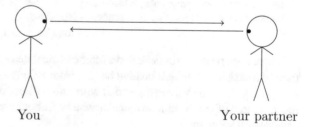

You think about your partner;

You think about what your partner thinks about you;

You think about what your partner thinks about what you think about your partner

⋮

If human interaction is this complicated, is there a way to analyze it in a consistent and logical manner? For a game theorist to overcome this difficulty, he/she must

Fig. 1.6 A game theorist and human interactions

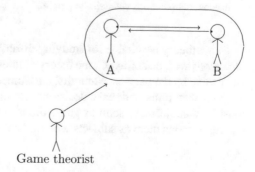

become an outside observer. Figure 1.6 illustrates this situation. However, human interactions are so complicated that it seems hopeless to decipher the infinite regress we have just seen.

The founders of game theory faced the exact same question. In their seminal book on game theory, von Neumann and Morgenstern wrote:

> First let us be aware that there exists at present no universal system of economic theory and that, if one should ever be developed, it will very probably not be during our lifetime. The reason for this is simply that economics is far too difficult a science to permit its construction rapidly, especially in view of the very limited knowledge and imperfect description of the facts with which economists are dealing. Only those who fail to appreciate this condition are likely to attempt the construction of universal systems. Even in sciences which are far more advanced than economics, like physics, there is no universal system available at present.
>
> —von Neumann and Morgenstern (1947, p. 2).

In order to proceed with the creation of a new theory analyzing human behavior, they emphasized the necessity of starting the analysis with a simple problem:

> It is necessary to begin with those problems which are described clearly, even if they should not be as important from any other point of view. It should be added, moreover, that a treatment of these manageable problems may lead to results which are already fairly well known, but the exact proofs may nevertheless be lacking.
>
> —von Neumann and Morgenstern (1947, p. 6).

There is no reason to believe that a mathematical theory is more suitable to analyze market transactions, which involve human interactions as a matter of course, than to analyze interactions between persons with and without disability. This is parallel to the argument of von Neumann and Morgenstern about the difference between natural science and economics:

> It is not that there exists any fundamental reason why mathematics should not be used in economics. The arguments often heard that because of the human element, of the psychological factors etc., or because there is–allegedly–no measurement of important factors, mathematics will find no application, can all be dismissed as utterly mistaken. Almost all these objections have been made, or might have been made, many centuries ago in fields where mathematics is now the chief instrument of analysis. This "might have been" is meant in the following sense: Let us try to imagine ourselves in the period which preceded the mathematical or almost mathematical phase of the development in physics, that is the 16th century, or in chemistry and biology, that is the 18th century. Taking for granted the skeptical attitude of those who object to mathematical economics in principle, the outlook in the physical and biological sciences at these early periods can hardly have been better than that in economics–*mutatis mutandis*–at present.
>
> —von Neumann and Morgenstern (1947, p. 3).

Game theory is suitable for studying disability-related issues from the viewpoint of persons with disability. Game theory studies the interactions of people who think and decide by themselves. Naturally, it assumes that their objects, human beings, are the ones who think and make decisions. No one should be a mere object; everyone must be a subject. As soon as game theory puts persons with disabilities into the picture, it treats them as subjects who make their own decisions.

The idea that everyone is a subject who makes his/her own decision is indeed what disability studies has been emphasizing. The slogan "nothing about us without us," which is the underlying theme for the United Nations Convention on the Rights of Persons with Disabilities, shares the same spirit.

Game theory also regards various phenomena as something that are determined by the interaction between human decisions and surrounding environments. It ascribes the outcome not only to individual traits, but also to the consequence of the interaction between them.

As a matter of course, various systems of the society, from the height of each step of stairs to the tax and social security systems, are made to fit the needs of "ordinary" people. This is true not only in Japan but also in many other countries. As a result, those who are out of the "ordinary" have difficulty in living through these systems. Although Japan, like many other developed countries, guarantees the basic human rights and the subsistence levels for living, these other people are classified into groups like homeless people, single mothers, the disabled, and so forth. Their basic needs are determined by the government, each specific need being labeled "luxury", and they have to live along as a social burden.

The purpose of our research project is to develop a new field of study on barriers in socio-economic contexts by introducing economics, especially game theory, into disability-related issues. Although this book does not present a general framework of disability, nor does it follow disability studies, it shares some of their motivation. Disability studies, first developed in the United Kingdom in the 1980s, is an academic field of study that regards disability as a social construct rather than something associated with individuals (see, e.g., Oliver 1984, 2013). It differentiates disability from impairment that is ascribed to individuals' physical and mental traits. Disability studies holds that it is disability, a social construct, rather than impairment, comprising individuals' traits, that induces a variety of social disadvantages to a specific group of people.

An economic viewpoint is essential for understanding and coping with these problems. Economics and game theory in particular have developed tools to study not only the market mechanism but also customs and institutions as endogenous and integral parts of economies. Economics and game theory investigate human behavior, shedding light on incentives and their interdependence. In particular, game theory is useful for this purpose. With these theoretical methods, we study the disabilities as endogenous institutions.[7]

---

[7] In the sequel, we often use the term "model." However, the term "model" used in the present analysis is the one used in game theory as opposed to the term used in disability studies such as a "social model of disability.".

1.4 Organization of the Book

The main body of this book consists of three parts. Part I draws the reader's attention to the study of economy and disability through some relevant cases. Part II introduces game theory and shows its basic use, especially in relation to the issues regarding disability. Simple mathematics is introduced and used in this part. Part III is related to more theoretical accounts of disability-related issues than those discussed in Part II.

Part I consists of a number of different cases. This part is divided into four chapters. Chapter 2 concerns the independent living of a person with disability and its relation to the market. Economics is a discipline that presumes independent individuals who participate in the market and make decisions by themselves. Persons with disability are often regarded as someone who must depend upon others to live. However, we regard them as market participants just like persons without disability. In fact, we would like to argue, through four cases, that their independent lives are enhanced through market activities. Nonetheless, the market is not well organized for persons with disability compared to those without disability. We argue, through a couple of cases, that it is not the market but rather a lack of market that prevents independent living for persons with disability.

Chapter 3 seeks the reason that persons with disability suffer. We often consider persons with disability as someone who needs extra care, and those without disability as someone who do not. In reality, however, persons without disability are someone who have been well taken care of by the society and the market. This chapter focuses on the minority and majority as the main source of difference between the two groups of people rather than those with and without disability.

Chapter 4 is concerned with prejudices induced by the dominant views of the society. The case of a person with color difference is a canonical case where the views of the majority are imposed upon the minority. The difference between this chapter and the previous one is that Chap. 3 is mainly concerned with behavior, while Chap. 4 is concerned with preferences.

Therefore, creation of markets for a minority is one of the critical issues. Chapter 5 considers three cases that have created the markets for minorities. The first case is a historical case that describes the development of an eyeglasses industry. In a society without eyeglasses, people with poor eyesight are considered as persons with disability. In such a society, visually impaired people were sometimes regarded as having an intellectual disability. The second case is the art market. Here, we observe how an art market has been created for the artists with intellectual disabilities. The third case, which is not related to disability per se, concerns *juku*, or private classes after school. Although one might call them "cramming" schools, this term reflects only part of what *juku* means. In reality, *juku* in Japan has created a market for children who are not accommodated in school. This case shows how a child is accommodated by the society through *juku*.

Part II introduces game theory and shows its basic use, especially for the disability-related issues. Chapter 6 introduces game theory. Chapter 7 shows the basic analysis of how disability-related issues are explained in a game-theoretic framework. In particular, the relationship between the minority and the majority is studied.

Part III presents a full-fledged analysis of economy and disability. Although this analysis captures certain aspects of various disability-related issues, it is not confined to these issues but is applicable to many other societal issues. Thus, most of the analyses try to use general language in game theory as opposed to the one specific to disability.

Chapter 8 introduces a dynamic process called the best response dynamics. This is an evolutionary process that does not require full rationality on the game players. In societal phenomena like disability, the evolutionary game theory sometimes provides us with a more powerful tool than the standard game theory, which assumes full rationality of the players. There is a brief discussion on the difference between deterministic and stochastic processes, with the claim that the best response dynamics, one of the deterministic dynamic processes, is appropriate for the kind of analysis considered here.

Using the best response dynamics defined in Chaps. 8 and 9 examines how cooperation emerges through voice or cheap-talk, which consists of intrinsically payoff-irrelevant pieces of information. The same voice may have different meanings and effects in different games, i.e., different strategic situations.

Chapters 10 and 11 analyze the interaction of conventions, or different behavior patterns. Cases in Chap. 3 are related to the theories examined here. Chapter 10 presents a general framework to analyze the interaction, while Chap. 11 focuses more on minority-majority issues in the economic arena. Chapter 10 develops a two-society model to analyze the interaction of societies with different conventions. After characterizing various equilibria, we apply an evolutionary approach to this world to see what happens if the two societies are integrated over time. For example, if the two societies are of similar size, the integration results in an "eclectic" convention, which leads to Pareto improvement. However, if one society is significantly smaller than the other, the smaller society whose convention disappears is worse off if its original convention is superior to that of the larger society. The two societies may be interpreted as two groups within a single society such as people with and without disability. Different groups have different traits or preferences so that optimal actions and desirable social outcomes may be different from each other.

Chapter 11 considers an economy in which coordinated interaction is required for transactions. Coordination on the language two parties use is a typical example. When there are two populations with different sizes and social backgrounds, it is shown that the smaller population (the minority) suffers from disadvantages in trade even if other physical conditions are symmetric between the two. In such a case, the minority may have to assimilate into the world of the majority to mitigate the disadvantages.

Chapter 12 is a new analysis partly inspired by disability-related issues. Indeed, the cases examined in Chap. 4 are related to the theory developed in Chap. 12 and in Chaps. 13 and 14. Standard game theory takes a model, or the structure of a game, as given and applies a solution concept such as Nash equilibrium or a behavior rule to the model in order to derive the strategies/behavior of the players of the game. In contrast, the theory of Chap. 12 takes experiences, or chunks of impressions, as primitives, which the agent then uses to construct a model. Some axioms are used in constructing a model. In the sense that a general structure of the game is induced based on limited experiences, the present analysis calls this theory *inductive game theory*.

Chapters 13 and 14 analyze segregation, discrimination, and prejudices, which are the traits that persons with disabilities have been enduring for a long time. The market for "ordinary" people and the market for persons with disabilities are often segregated. For example, the labor market in Japan has a quota system for the employment of persons with disability. While this is an effective policy to increase the employment of persons with disability, it has created a segregated market for them. The analysis uses an anecdote of multiple festivals that people might attend. In a segregation equilibrium, two groups of people go to different festival locations, just like persons with disabilities go to a specific labor market for employment rather than try to compete with persons without disability. Through these experiences, prejudices may emerge. We consider ways to eliminate prejudices, and, in doing so, we adopt a new approach called inductive game theory, which is discussed in detail in Chaps. 12, 13 and 14.

Chapter 15 constructs a model that studies the case of hereditary deafness on Martha's Vineyard Island in Massachusetts, U.S.A. in past centuries, where the island community adjusted itself to the hereditary deafness and made it a "non-disability."

Throughout the analysis, we agree with disability studies researchers who have asserted that disability presents not only physical/medical issues, but societal/economic issues, and that the notion of disability is socially relative. However, the field of disability studies does not have formal language that expresses the source of the problems and analyzes the society in a logically coherent and precise manner. We propose that a game-theoretic analysis should be used to show its usefulness in understanding disability-related issues.

It is worth mentioning two things that are *not* covered in this book. First, this book does not relate the game-theoretic analysis to disability studies.[8] Although the contribution of disability studies to the empowerment of persons with disability should not be under-valued, its method is very different from the present analysis as British disability studies follows Marxian traits when it comes to a discussion of economic issues. Discussing disability studies in relation to economics in this book is like writing a book on Karl Marx and John F. Nash. I would like to entrust such a task to more able and ambitious researchers than I.

Second, we do not conduct policy analysis. The analysis of policy making is one of the most important tasks. This book focuses on the essence of disability through the eyes of game theory, trimming the details that are not necessarily common to various phenomena.

---

[8]The reader may refer to Sugino (2007), which covers both English and American disability studies. See also Kawashima (2011) for the difference between the two streams of the concept of disability. According to Kawashima (2011), the social model of the United Kingdom is that "it adopts the distinction between 'impairment' and 'disability' as proposed by UPIAS (Oliver 1996)," while Americans with Disabilities Act (ADA) defines the term "disability" as "a physical or mental impairment that substantially limits one or more major life activities," which emphasizes the interaction of impairment and disability.

Part I
Disability and the Market: Cases

Chapter 2
The Market for All?

2.1 Independence of Persons with Disability

Professor Shin-ichiro Kumagaya is an Associate Professor at the University of Tokyo, and is well known for his self-studies of disability. He has suffered from cerebral palsy for his entire life after enduring a high fever immediately after his birth. He is currently using a wheelchair.

Professor Kumagaya's mother devoted her life to his education. As instructed by a doctor, she demanded that he moved his hands as well as other children without disability, and she sometimes scolded him when he could not perform the task. His mother was a good carer, and, when in high school, Prof. Kumagaya worried that he would die if his mother died because he was so dependent on her.

Professor Kumagaya moved to Tokyo from Yamaguchi Prefecture, located at the western edge of mainland Japan, at the age of 18 when he became a college student, leaving his mother behind. He thought, "I will become independent." Of course, his independence may be different from that of people without disability because he still needed a lot of support to do daily things. A question then arises as to what made him think he became independent.

The term "independence" is an antonym of "dependence." Professor Kumagaya stopped depending on his mother and started fabricating his life on his own. Today, his assistants number more than 30, and although he asks them for assistance, he does not depend on one particular person. He used to depend upon a thick lifeline (his mother), but now, he ensures that he is supported by a web of thin but many lifelines.

Professor Kumagaya's case reminds me of the Japanese novel called "The Bears of Mt. *Nametoko*" by Kenji Miyazawa, a famous writer of many fables. In this novel, a hunter called Kojuro lived in the mountain and knew his animals well. He felt awful whenever he killed a bear, but it was necessary if he were to make a living. A brave Kojuro had to sell the bear's skin to a merchant in the town at an outrageously low price. That such a brave man like Kojuro had to beg the merchant for trade was a sad part of the story.

© Springer Nature Singapore Pte Ltd. 2019
A. Matsui, *Economy and Disability*, Economy and Social Inclusion,
https://doi.org/10.1007/978-981-13-7623-8_2

While some readers may get angry at the absurdity of the market, others will understand that little can be done because that is how the market operates. At this point, we should pause to consider what these opinions may imply.

The central problem is that Kojuro did not know any other way to exchange the skin for money except selling it to the merchant. Kojuro's life was dependent upon this particular merchant. Although the scenarios of a mother's affection and a merchant's greed are totally different, the two situations are similar in the sense that one agent depends on the other.

The merchant's greed caused a problem not because there was a market, but because there was a *lack* of market, and Kojuro did not have an alternative. To Kojuro, the merchant was a lifeline, and the merchant knew it and took advantage of Kojuro.

For better or worse, the characteristics of the market lie in the lack of restrictions. In the market economy, we, as a matter of course, depend upon many things. Without supermarkets and/or grocery stores, we cannot eat our everyday meals, but stores also depend upon their customers In spite of these dependencies, we do not strongly depend upon a particular store or person. If we cannot buy something in one store, we can go to the next store. Even if one customer does not like your store, other customers will buy things from your store. The essence of the market economy lies in the life supported by loose but many ties.

Professor Kumagaya says, "An independent life is obtained when you can depend upon many things so that you feel like you are not dependent on one particular thing." Independence is the situation where you depend upon various things.

Those who are obsessed with making and saving money are not independent people. From the present viewpoint, they show an addictive tendency and are overly dependent on money. Such people are so anxious about earning money that they cannot do anything without performing some calculation to consider their actions in terms of money.

Although we all need money in our daily life, there are many things we cannot simply buy. Mental satisfaction is often something that cannot be obtained in a market. Parents' affection for their children can never be bought in a market.

We must also recognize that depending upon the market itself is a fragile situation and we must find ways to deal with the market when things go wrong. After the earthquake that hit eastern Japan on March 11, 2011, the market collapsed, and people in the Tohoku region had to find ways to obtain goods through different channels.

Sometimes, the market fails because some people fall into the consumer's trap of addiction. Knowing this, some firms try to develop products to which consumers will become addicted. To prevent addiction, the government sometimes regulates the market. The use of drugs for psychotropic rather than medical purposes is a typical example. Opium, cocaine, and heroine are among them. Caffeine, nicotine, and alcohol are also drugs. The habitual and excessive intake of coffee, tobacco, and/or alcoholic beverages are examples of drug dependence and addiction.

To fight against drugs, there are basically two methods.[1] The difference between the two reflects the two views of drug addiction. One view is that drug addiction is an illness, while the other considers it to be a crime.

If the government views addiction as an illness rather than a crime, the prescription of drugs is often allowed. For example, to reduce dependence on drugs, in 1967 the British government allowed the prescription of drugs to addicts at drug treatment clinics. Under this system, some addicts can obtain legal supplies from the clinic, while others can obtain supplies from a neighborhood pharmacy and medicate themselves. These clinics also provide the addict with social services such as psychotherapy. As a result, a large proportion of the addicts have become productive, socially useful members of the community.

On the other hand, in countries where the addict is treated as a criminal, physicians are sometimes prevented from prescribing opiates to the addict. Acceptable treatments are to put the addict in an institution for months with strict regulation against ambulatory care until the person is drug-free. Drugs are totally prohibited even under a physician's care. Estimates of cures based upon decades of such government-regulated procedures range from 1 to 15%.

In addition to drugs, there are many other products or concepts that may cause addiction, but it is either impossible or inappropriate for the government to regulate the products. According to "The Fix" by Damian Thompson, the iPhone is one of such examples. He indicates that Apple actually tries to make the iPhone "addictive":

> Former Apple executives who frequently brief American technology blogs off the record about the internal culture at Apple's headquarters in Cupertino, California describe the lengths the organisation goes to in order to create coveted products. There's a design-dominated power structure that results in hushed reverence when Jonathan Ive, Senior Vice President of Industrial Design, walks into the boardroom. "Marketing and design have been fused into a single discipline at Apple," says Yiannopoulos. "Everything, from product strategy to research and development, is subordinate to making the products as beautiful and compulsive that is, as addictive as possible." (The Fix, p. 19)

In some extreme cases, iPhones have created persons with disabilities, especially mental disabilities:

> (Their strategy) works. To quote an extreme example, in 2010 a schoolboy in Taiwan was diagnosed with IAD iPhone Addiction Disorder. According to Dr Tsung-tsai Yang of the Cardinal Tien Hospital, his eyes were glued to his phone screen all day and all night. Eventually, "the boy had to be hospitalised in a mental ward after his daily life was thrown into complete disarray by his iPhone addiction". (The Fix, pp. 19–20)

To recover from addiction, one should gradually depend upon other products or relationships rather than the one that causes the addiction. Once again, everyone should depend upon many things, therefore creating a network of products or services upon which the patient can depend. This concept is more important than removing the source of addiction.

---

[1] This part of description is based on Steiner (2016).

To summarize, for all of us to live an independent life, persons with and without disability should depend upon many things rather than one thing. The market is a major instrument that can help us achieve this goal.

2.2 Supporting Others While Being Supported

In 2016, the author visited the Taira Special Needs Education School for handicapped children in Iwaki-City, Fukushima, Japan. Iwaki is the second largest city in the entire Tohoku region and is located in the south-east corner of Fukushima Prefecture, facing the Pacific Ocean. It was badly affected by the Great East Japan Earthquake in 2011.

The school is situated in a hilly part of the city. The main building of the school receives plentiful sunshine and has a large atrium with an even larger ramp that connects the first and second floors so that wheelchair users can easily move between the two floors.

The purpose of the visit was to meet student members of a volunteer activity club, who won the first prize in the contest "The Second Fukushima High School Contest for Social Activities."

Mr. Miura, a freshman of the high school division, presented their activities on a campaign for raising funds. He said, "We, the persons with disability, have always been supported. We receive support as a matter of course. However, my *senpai* (senior student) who started these activities made me think that I can be the one that supports others while being supported by others."

After conducting a tour of the facility, Ms. Yukiko Aoki, the teacher of Mr. Miura, described how the volunteer activity club started. According to her, one student, Mr. Kazuma Sakamoto, had a passion to form such a club, which commonly existed in other schools. Kazuma commuted to school from the hospital next to the school. His home was destroyed by the earthquake in 2011, and his parents lived in a temporary house. It was difficult for them to live together.

After much consideration, Mr. Sakamoto came up with the idea of creating a volunteer activity club. He said, "Persons with disability have always been supported. I'd like to support others from now on." Having graduated, he now lives by himself with support, working for a company at home.

His passion was passed on to the younger generation. Ms. Yuki Higuchi, a junior of the high school division, said, "I joined the club because I wanted to support others rather than being supported. Activities to promote blood donation on the street were particularly hard but rewarding." In the beginning, few people bothered to accept pocket tissues with printed advertising for blood donation. However, Ms. Higuchi changed her method of asking people, and she was able to effectively distribute the tissues and the associated community service message. The key factor, she said, is "to speak out with a smile on your face." Her future aspiration was to obtain a college qualification in health and welfare to become a certified social worker.

The author invited students of the Taira school to the University of Tokyo in 2018 to see what happened to the volunteer activity club after that. *Kohai* (junior students) of Ms. Higuchi continued on with the club activities. Their presentation was entitled "# with". Ms. Rimi Ohmura explained that it is important to change from "support for persons with disability, support for elderly people" to "support *with* persons with disability, support *with* elderly people."

Mr. Kodai Tani, a wheelchair user, provided us with an interesting case. One day, he found that he was unable to climb up a wheelchair ramp even though he normally had no such trouble. The reason was that the ramp was too steep. "Why did it happen?" he asked the audience. "It happened since the ramp was made without the support of a wheelchair user." He highlighted this as an example where support should be provided not *for* persons with disability, but *with* them.

Mr. Tani then made a simple clinometer, an instrument that measures slope angle. He found that he could not climb a slope of 8°, while the other ramps that he normally used were 4° or 5°. Mr. Tani then made more than 50 clinometers and distributed them to attendees at an annual nationwide meeting for providers of volunteer works. He has since obtained some feedback.

Support for persons with disability may often end up as a situation where those with disabilities find themselves under coercion on them by those without disability. Such a scenario is like a planned economy where producers or a central authority decide what people should consume and force them to buy it. The producers in the planned economy care about how many products they *produce*. Supporters for persons with disability care about how many cases they file. On the other hand, in the market economy, producers have to consider how many products they can *sell*. To sell their products, they have to think carefully about what the consumers need. Consumer sovereignty is naturally embedded in the market mechanism. The same principle of sovereignty of persons with disability should be embedded in the support mechanism. The move from a placement-based system to a contract-based system is an important step for this change.

The key phrase of the Conventions on the Right of Persons with Disabilities (CRPD) of the United Nations is "Nothing about us without us." The students of Taira school put these words into practice.

Chapter 3
Economies for the Majority

Our society is generally designed for "ordinary" people. As a result, it tends to exclude people who are not sufficiently accommodated. The ordinary people often constitute majority, while those who are excluded become the minority. This chapter discusses three cases to illustrate how the "minority" suffers in our society. This chapter is also related to the subsequent theoretical chapters, especially Chaps. 10 and 11, which study these phenomena in detail.

3.1 Children with Hearing Disability

Welfare programs should be utilized not only for helping the weak, but also for removing the ceiling that prevents children with disabilities from growing. The idea of investment, rather than that of traditional care, is needed for them. While considering such an approach in 2013, the author visited Kanamachi-gakuen, which is a welfare facility that accommodates children with hearing disabilities.

Kanamachi-gakuen was shaken by the misconduct of a staff member more than 10 years earlier. The organization completely reformed their system and invited the school master of a school for the deaf to assume control as Director of the facility. Kanamachi-gakuen then utilized the Services and Support for Persons with Disabilities Act and started a new program.

About 20 children with hearing disabilities lived in Kanamachi-gakuen at the time of the author's visit. It was a Sunday, and we joined a Sunday school called "NPO Otsuka Club Kanamachi Learning School." Volunteers visited each Sunday to assist the children with their study. In addition to Mr. Yukio Morimoto, the Director of the school as well as a sign language interpreter, a college professor and students helped children of elementary school age to high school age with their studies. Among the children, some were studying for college entrance examinations. Some teachers were also deaf.

© Springer Nature Singapore Pte Ltd. 2019
A. Matsui, *Economy and Disability*, Economy and Social Inclusion,
https://doi.org/10.1007/978-981-13-7623-8_3

Mr. Kan Sugawara, a student of Tokyo Science University, had been at the school for 2 years. He told me that some teachers at special schools could not understand sign language, and that most of the children have to repeat the same materials at the Sunday school. He said, "We need this type of environment."

Some children enter Kanamachi-gakuen from outside the Tokyo metropolitan area. A student who successfully passed the entrance exam came from Fukuoka Prefecture in the Kyushu region, a western part of Japan. He had almost abandoned thoughts of going to college when Ms. Kumiko Hamazaki, the Director at the time, organized a camp to promote Kanamachi-gakuen. He said, "Most of the senior students who wanted to advance to college went to Tokyo. So, I wanted to go there and study hard." Although it was hard to commute to school, which was a 90-min trip one way, he studied hard to fulfill his dream.

The welfare programs of Japan were drastically changed in the 2000s after substantial reform. Before the reform, the welfare system was placement-based, where bureaucrats decided which individuals with a disability should be offered a place. After the reform, the system became contract-based, where persons with disabilities formed contracts with a facility, and the government subsidized 90% or more of the program. Upon this reform, Kanamachi-gakuen changed its policy of accommodating only children that were unable to be raised at home by their parents, and started accepting other children on a contract basis.

Welfare programs try to secure the bottom line, and therefore, they do not lead to the idea of raising leading figures. Children with hearing disabilities must be exposed to wider society in the same way that children without disability are exposed. If we label such children the weak and provide them with an education from the viewpoint of welfare programs, then we will stunt their development. Mr. Sugawara said, "Basically, special supporting schools aim at the bottom level students. Although we need the education that normal schools provide, it is difficult at this point."

At high schools, those without disability improve themselves through friendly rivalry. However, students with hearing disability can enter schools for the Deaf without any effort, and therefore, they do not have much incentive to work hard. As a result, they do not try to improve themselves, which makes it difficult to maintain their learning ability.

Ms. Hamazaki said, "For children with hearing disabilities to spend their social lives at the same level as those without disability, they need to be able to use Japanese as well as those without. A child with a disability tends to rely on other children without disabilities when they are together. Watching such situations, children will follow the same trail. To break such a negative chain, staff members try to cultivate the initiative of children with disabilities and show them situations where both children with and without disabilities work on an equal basis."

After entering a college, the next task that awaits a deaf student is how to "listen" to the lectures. Once again, the efforts of the pioneers were impressive. For example, Mr. Sugawara established a team of sign language interpreters and note-takers. The college responded to his effort. The college that accepted the student from Fukuoka started preparing to assist him.

The spirit of independence cannot be cultivated without such steady efforts. Ms. Hamazaki said, "One day, I wish to see a deaf person become the Director of this facility." Hopefully, this sentiment will be the basis of the future development of policy toward disability in Japan.

Another notable observation at Kanamachi-gakuen was that everyone there "speaks" sign language. It was reminiscent of the book entitled "Everyone here spoke sign language (Groce 1985)". The book is about hereditary deafness on Martha's Vineyard, an island off the Massachusetts coast. The remarkable thing on the island was that from the Seventeenth century to the early Twentieth century, the rate of deafness was so high that both deaf and hearing islanders used sign language so that deaf people were not isolated from the community.

Kanamachi-gakuen strives to create a similar environment. Outsiders at Kanamachi-gakuen need an interpreter rather than the children. Lunch time was very difficult for the non-deaf. The children spoke sign language, and from time to time the person who spoke oral language kindly translated what they were saying. Despite this help, those without sign language felt isolated and were definitely in the minority.

The matter of who are in the minority and who are in the majority is determined by the fractions of people types in the society. In Kanamachi-gakuen, disability was determined by the society because physical impairment, the fact that one cannot hear, is not a significant issue compared with social disability (unable to communicate), which is a significant issue.

3.2 The Dancer Fighting with a "Forgotten Cancer"

Ms. Yurie Yoshino was a dancer gifted with both intelligence and beauty.[1] It is regrettable to use the past tense because she passed away in July, 2016, at the age of 48. She was a member of our team on "Research on Economy And Social Exclusion (REASE)."[2] She had fought for more than ten years against the "forgotten cancer" for which the five year survival rate was said to be 7%. She endured 19 operations and six radiation therapies.

Yurie was multi-talented and won the title of Miss Japan when she was a student at the University of Tsukuba. After graduation, she became a professional competitive dancer. At the Blackpool Dance Festival 2000, which is the most prestigious dance competition of the world, the pair of Yurie and her partner reached the last 24.

---

[1] The description of this section is partly based on Yoshino (2016).
[2] REASE is the next generation research project after READ.

After retiring from professional dance, Ms. Yoshino spent busy days as a dance teacher and volunteer. At this time, she noticed something wrong with her body and sought medical advice. She was diagnosed with ovarian cystoma, which is a benign tumor, and underwent surgery for its removal. However, after pathology testing, the tumor was identified as sarcoma, which is a malignant tumor. Suddenly, it was too late as the tumor that was initially considered benign has spread over her entire body.

The incidence of sarcoma in Japan is about 3000 out of more than 120 million people, or 0.0025%. Because sarcoma is uncommon, few doctors specialize in its treatment, and, therefore, there is little progress in research. In addition, sarcoma is easily confused with other diseases. In contrast, about one million new patients are diagnosed with cancer in general, and about four hundred thousand people die of cancer each year. Naturally, the Japanese Government focuses on cancers like stomach and lung cancers, which are more common. Dr. Masahiro Kami, Director of the Medical Governance Research Institute, a non-profit organization for which Ms. Yoshino worked, commented, "One cannot make a living if one specializes in Sarcoma only."

Sarcoma is classified as a rare disease. However, according to the World Health Organization (WHO), there are 50,008,000 rare diseases identified worldwide, and about 1 out of 15 people have a rare disease of some sort. In Japan, this equates to about eight million people suffering from rare diseases. This number exceeds the number of patients with cancer or diabetes. However, only 330 diseases designated as rare diseases are eligible for subsidies from the Japanese Government as of April 2017. The number of the patients covered by the system is about 1.5 million as estimated by the Ministry of Health, Labour, and Welfare. To be a designated rare disease, the disease must meet several strict conditions. First, fewer than 0.1% of the population suffer from the disease; second, a diagnostic method is established; third, the method of treatment is unknown. In particular, the second and the third conditions together narrow the window for designation. Because of these stringent conditions, the health system does not shed light on most rare diseases.

Yurie made an effort to establish a research-treatment center for sarcoma. In 2012, Dr. Ryosuke Tsuchiya, a board member of the Ariake Hospital Cancer Institute, was moved by her effort and founded the first "Sarcoma Center" in Japan. After this news was broadcast on a television program, the recorded number of cases of sarcoma doubled, and the research foundation was established.

Essentially, the Sarcoma Center was founded as the result of one patient's effort that moved conscientious doctors. While it is important to lobby the government for policy reform on rare diseases, Yurie Yoshino showed us that sometimes it is effective to appeal directly to the medical community. Yurie wrote a book (Yoshino 2016) about her fight against the forgotten cancer. It was finally published and delivered to her just the day before she passed away. I pray for the repose of her soul.

3.3 Chronic Myelogenous Leukaemia

Ms. Yuko Kodama, a nurse and a researcher at the University of Tokyo as well as a member of READ, was shocked when she heard that one of her patients refused to be treated with a drug.[3] The reason, the patient's family explained, was the cost of the drug. The patient was suffering from chronic myelogenous leukemia (CML), which is acknowledged as a rare disease in Japan.

According to Desforges et al. (2018), "Chronic myelogenous leukemia (CML) is characterized by the appearance in the blood of large numbers of immature white blood cells of the myelogenous series in the stage following the myeloblast, namely, myelocytes. ... The disease is most commonly encountered in persons between 30 and 60 years of age."

Imatinib, also known commercially as Gleevec, was approved in U.S.A. in 2001 for the treatment of CML. Imatinib is designed to target abnormal proteins that breed cancer cells. Imatinib is different from traditional anticancer drugs such as interferon, which do not differentiate between cancer cells and normal cells. As a result, imatinib is highly effective; the five-year survival rate for previously untreated CML patients is nearly 90%. Its success set an important milestone in cancer treatment since targeted therapy is more effective and safer than traditional one. The development of imatinib for the treatment of CML has become a model of the subsequent development of other targeting therapies.

Japan normally has slow approval processes for the use of new drugs. However, in the case of imatinib, the government permitted the importation of imatinib in 2001, taking into account that it had undergone intensive testing abroad, and that it was intended for severe but rare diseases like CML.

Imatinib was very effective and safe, but it had one critical problem: it was expensive. At the time of our research around 2009, 400 mg of imatinib, the standard daily dose, cost about 14,000 yen, or 150 USD (using the average exchange rate of 104 yen per dollar). Thanks to the medical insurance system, the patient does not have to pay full cost (otherwise, it would be impossible for most people). For example, if the patient is under 70 years old and in the middle income bracket, and if he or she takes the same dose as above for 30 days, then the patient pays 81,630 yen, or about 780 USD.

However, the actual practices are different across hospitals and clinics. Our clinic initiated research into the financial burden of imatinib upon patients with CML in a study coordinated by Ms. Kodama, with assistance from Ryoko Morozumi to cover economic aspects.

The research method was to distribute a questionnaire to 1200 patients between May and August, 2009. We retrospectively surveyed their household incomes, out-of-pocket medical expenses, final co-payments, and the perceived financial burden of their medical expenses in 2000, 2005, and 2008.

A total of 577 patients completed the questionnaire. The median age was 61 years. Out of 534 valid responses, 229 (42.9%) patients were working full-time, and 226

---

[3]This section is partly based on Kodama et al. (2012).

(42.3%) were either retired or not working. The median household income was 4.1 million yen (40,000 USD). The final co-payment in 2008 was 370,000 yen (3600 USD). A simple calculation shows that about 9% of household income was spent on medical expenses. For the low-income cohort with an annual income of less than 2.7 million yen (26,000 USD), the median ratio of co-payment to income was close to 20%.

A major problem of targeted therapy with imatinib relates to the financial burden on patients. Treatment of CML must last for life; otherwise, the disease progresses and ultimately reaches the stage where treatment is no longer effective.[4] For example, if a 40-year-old woman is diagnosed with CML and manages to live to the average life expectancy in Japan (87 years), the co-payment of medical expenses including targeted therapy with imatinib is assumed to be 370,000 yen, which equates to a total payment of more than 17 million yen, or 180,000 USD.

This problem is not unique to the treatment of CML. Many rare diseases are chronic and may last a lifetime.

The medical insurance system of Japan does not consider the fact that there are chronic diseases that last for many years and incur very high expenses. Payment is determined on a yearly basis, and the payment increases linearly as time goes on, even though the annual payment is capped. This leads to a heavy financial burden on patients, and especially on low-income families.

If the medical insurance of Japan is "insurance" in its true sense, then it should prepare for expenses of 13 million yen rather than 370,000 yen. The total welfare increases if the system is modified in such a way that the total lifetime co-payment is reduced. To calculate the cost of risk, let us simply compare the current system with the optimal system with complete risk hedging, and ignoring various incentive problems. With some other assumptions, the *ex ante* lifetime welfare loss of individuals is estimated.

According to the estimation presented in the Appendix to this section, the lifetime welfare loss ranges from about 850,000 o about 1.5 million yen, depending on the income level and risk attitude. The total lifetime welfare loss caused by the incomplete insurance system relative to a complete one was obtained by multiplying the above numbers by 120 million, the total population of Japan, and amounts to:

$$100\text{–}180 \text{ trillion yen,}$$

with an annual loss of

$$1.7\text{–}3 \text{ trillion yen.}$$

This estimation ignores, among other factors, the side effects of the treatment, a decrease in income, and an increase in living costs caused by suffering from a rare disease. These particular factors are all detrimental to the welfare of the patient in question.

---

[4]The subsequent analysis, including the following appendix, is independent of Kodama et al. (2012).

Appendix: Simple Simulation for the Welfare Loss Caused by Incomplete National Insurance System

This appendix presents a simple simulation to calculate the cost of rare diseases under the current insurance system and other insurance systems.

Consider a business person, who is assumed to be male, and works for 40 years from 21 to 60 years old. After retirement (with no retirement allowance), he lives on a pension for 20 years until age 80 with the annual pension of I million yen. It is assumed that necessary expenses during working age would be higher than after retirement. Such expenses would include costs of raising children and mortgage or rental payments. To reflect this and to simplify the analysis, we assume that only I million yen is at his disposal so that from age 21 to 80, the subject consumes as if his income were flat at I million yen.

Suppose that a person has an annual utility function with a constant relative risk aversion. Then the utility function $u(x)$ of this person can be represented as:

$$u(x) = \frac{x^{1-\sigma}}{1-\sigma}, \quad (\sigma \neq 1),$$

where x is the amount of consumption in million yen beyond the necessary expenses mentioned above, and $\sigma = -\frac{u''(x)x}{u'(x)}$ is the constant rate of relative risk aversion that measures this person's risk attitude. Furthermore, this person tries to maximize the discounted sum of utility:

$$U(x_{21}, x_{22}, x_{23}, \ldots, x_{80}) = u(x_{21}) + \delta u(x_{22}) + \delta^2 u(x_{23}) + \cdots + \delta^{59} u(x_{80}).$$

subject to the budget constraint (with no liquidity constraint) given by

$$x_{21} + \delta x_{22} + \delta^2 x_{23} + \cdots + \delta^{59} x_{80}$$

$$\leq LI \equiv I + I\delta + I\delta^2 + \cdots + I\delta^{59} = \frac{I(1 - \delta^{60})}{1 - \delta},$$

where LI is the lifetime income, or the discounted sum of total income. It is verified that the best consumption stream of this person is $(I, I, \ldots, I)$, i.e., the one in which he consumes I million yen every year. The total utility U_N is

$$U_N = u(I)\frac{(1 - \delta^{60})}{1 - \delta} = \frac{I^{1-\sigma}}{1 - \sigma}\frac{(1 - \delta^{60})}{1 - \delta},$$

where, in notation, U_N stands for utility with no disease.

Suppose that there is a possibility that the subject suffers from a rare disease at the age of 40. As a result of ideal treatment, this person lives to age 80. The probability of this event (suffering from a rare disease) is $p\%$. The subject does not consider this possibility until he actually suffers from the disease. With an annual cost of treatment of C million yen per year, the total utility becomes

$$U_D = u(I)\frac{(1 - \delta^{20})}{1 - \delta} + u(I - C)\frac{(\delta^{20} - \delta^{60})}{1 - \delta}$$
$$= \frac{I^{1-\sigma}}{1 - \sigma}\frac{(1 - \delta^{20})}{1 - \delta} + \frac{(I - C)^{1-\sigma}}{1 - \sigma}\frac{(\delta^{20} - \delta^{60})}{1 - \delta},$$

where U_D stands for utility with a disease.

Then the expected utility is

$$EU = (1 - p)U_N + pU_D.$$

It is clear that the optimal policy toward rare diseases is to be fully insured. The question that remains is the magnitude of the welfare loss measured in terms of expenses if the current policy is maintained. This calculation requires quantification of the certainty equivalence (CE) of the situation with the possibility of disease without full insurance:

$$CE = \left[(1 - \sigma)\frac{1 - \delta}{1 - \delta^{60}} EU \right]^{\frac{1}{1-\sigma}}.$$

Thus, an individual welfare cost measured in terms of income caused by an incomplete insurance system (relative to a complete insurance system) is given by

$$LI - CE.$$

Then the welfare cost evaluated at the age of 40 becomes

$$(LI - CE)/\delta^{20}.$$

This value is listed in Tables 3.1 and 3.2.

Explanation of the choices of variables are as follows. The income is a hypothetical one, but it is assumed that the utility function takes the value beyond a reference point. For example, if a person earns 5 million yen per year with some necessary expenses, such as education costs and housing loan, of 3 million yen, then the person's utility comes from the remaining 2 million yen. The rates of relative risk aversion are

Table 3.1 Lifetime loss, simulation 1: $I = 2, C = 0.37, p = 0.067$

(loss in thousand yen)				
δ	0.99	0.99	0.999	0.999
σ	0.5	2	0.5	2
Welfare loss	857	993	1015	1176

Table 3.2 Lifetime loss, simulation 1: $I = 1$, $C = 0.37$, $p = 0.067$

(loss in thousand yen)

δ	0.99	0.99	0.999	0.999
σ	0.5	2	0.5	2
Welfare loss	907	1266	1074	1497

roughly the lower bound and the upper bound taken from empirical researches (see, e.g., Shimono 2000). The discount factor is small, reflecting the recent low interest rate in Japan. The probability of suffering from a rare disease is taken from the fraction of patients with rare diseases as reported by WHO.[5]

---

[5] http://www.who.int/medicines/areas/priority_medicines/BP6_19Rare.pdf.

Chapter 4
Dominant Views and Prejudices

4.1 "There Is No Such Color!"

Our research group, Research on Economy And Disability (READ), posted a website on color perception.[1] It shows how signs look different if they are adjusted for different groups of people in terms of color perception. Various signs are shown on the page, including traffic light signals for pedestrians and signs for toilets.

In Japan, as in many other countries, the color perception of most people is type C. However, there is a small fraction of persons with different types of color perception. In fact, 5% of the entire male population are of type D or P, who cannot distinguish green and red. The traffic light of our webpage is designed in such a way that persons with different color perception tend to interpret the sign in a different manner.

Our traffic lights utilize differences in both hue and saturation. Respondents with type C color perception tend to pay attention to hue and distinguish green and red so that one sign appears to them as "Walk!" While respondents with type D or P color perception, who you cannot distinguish green and red, pay more attention to saturation better than type C respondents so that the same sign appears to them as "Stop!"

After we posted our webpage, we had a response from a man with "colorblind," who we refer to as *Jun*. Jun readily agreed to publish his response[2]:

> Art class was difficult for me to in elementary school. ... I was told repeatedly, "There is no such color!" or "Its color is not like that!" Talking about color was taboo at home, too. ... These situations gave me strong negative feelings. However, when I was in ninth grade, I met a math teacher who was also colorblind. The teacher said to me, "I am colorblind, too. About 5% of the population are colorblind, and it has a particularly high proportion among intellectuals." ... He gave me an opportunity to see color blindness in a relative manner. Until that moment, I thought of color blindness as a disadvantage one could never overcome. However, now I come to think of it as a trivial inconvenience.

---

[1]The web page is in Japanese with http://www.rease.e.u-tokyo.ac.jp/enlighten.html (as of May 7, 2018). The members of the team on this issue are Yoshiki Tomita, Taichi Niwa, and Hiromi Tojima. Akihiko Matsui is in charge of the web site.

[2]Translated by the author.

© Springer Nature Singapore Pte Ltd. 2019
A. Matsui, *Economy and Disability*, Economy and Social Inclusion,
https://doi.org/10.1007/978-981-13-7623-8_4

Previously, many jobs were not legally available to people with color blindness. Although this legal restriction has been removed, society has not changed to accommodate people with this condition. For example, the traffic signals for "Go" and "Stop", are still distinguished only by green and red colors. The purpose of constructing our webpage is to point out this problem.

According to Nassau (2018), *colour* (color) is described as follows:

> Colours result from the electromagnetic radiation of a range of wavelengths that are visible to the eye. The three characteristics of hue, saturation, and brightness are commonly used to distinguish one colour from another.

—"Colour," Nassau (2018)

Logically speaking, these three characteristics, hue, saturation, and brightness, are artificial. However, we need to understand these traits if we wish to better understand their association with disability:

> The hue is that aspect of colour usually associated with terms such as red, orange, yellow, and so forth. Saturation (also known as chroma or tone) refers to relative purity. When a pure, vivid, strong shade of red is mixed with a variable amount of white, weaker or paler reds are produced, each having the same hue but a different saturation. These paler colours are called unsaturated colours. Finally, light of any given combination of hue and saturation can have a variable brightness (also called intensity or value), which depends on the total amount of light energy present (Nassau 2018).

The next question is how color is perceived by human eyes and brains. The most well-known theory is the trichromatic theory. This theory postulates three types of color receptors, known as cones, in the eye. The three types of cones have sensitivities for blue, green, and red. These three sets are often designated as S, M, and L for their sensitivity to short, medium, and long wavelengths.

One of the strengths of the trichromatic theory is that the existence of some color blindness can be explained as a lack of function of one or more sets of the cones. For example, persons with deuteranopia (M set missing) or protanopia (L set missing) perceive only blue and yellow. People who have no functioning cone set are extremely rare and can perceive only greys.

As mentioned above, about 5% of the male population (and 0.2% of the female population) have either deuteranopia (type D) or protanopia (type P).[3] Currently, they are referred to as "colorblind." However, most difficulties that this minority of the population face arise because society is dominated by type C people, who tend to be the ones that make the rules of the society. In other words, only type C people are fully included in the society.

As our homepage suggests, if our society were to be designed only for type P and type D people, then type C people would become "colorblind." It is society, rather than nature, that creates a division between the majority and the minority.

---

[3]The difference in ratio between male and female is caused by genetics. Type D gene is a recessive gene. For men, the genotype of the mother is critical. A man has a 50% chance of color blindness if his mother has this gene. A man has a 100% chance of color blindness if his mother is colorblind. A woman can only be colorblind if her father is colorblind and the mother gives her this particular gene.

Sacks (1996) wrote about a unique island called Pingelap.[4] Pingelap is one of tiny atolls scattered in the Pacific Ocean, part of Pohnpei State of the Federated States of Micronesia. What makes Pingelap unique is that about 8% of the total population of the island are achromatopic (completely color-blind). Complete color-blindness is rare even compared with deuteranopia and protanopia. According to Sacks (1996), only one person in 30,000 or 40,000 people is completely color-blind.

In Pingelap, Sacks and his friends met their interpreter, James, who was also achromatopic. Bob, a friend of Sacks, wondered how achromatopic people could tell if, say, bananas are ripe:

'But what about bananas, let's say–can you distinguish the yellow from the green ones?' Bob asked.

'Not always,' James replied. ' "Pale green" may look the same to me as "yellow".'

'How can you tell when a banana is ripe, then?'

James's answer was to go to a banana tree, and to come back with a carefully selected, bright green banana for Bob.

Bob peeled it; it peeled easily, to his surprise. He took a small bite of it, gingerly; then devoured the rest.

'You see,' said James, 'we don't just go by colour. We look, we feel, we smell, we *know* –we take everything into consideration, and you just take colour!'

The society that includes a minority must do so in a way that the majority feel comfortable. The majority-minority concept tends to divide society and creates unnecessary prejudices. We all try to adapt to society one way or the other. However, if too many barriers are placed within our society, those who are not "ordinary" become a minority and suffer from discrimination and prejudice.

Let me conclude the story of Jun. He sent me a drawing of the islands of the Seto Inland Sea where he lives, and he also sent me a photograph that was taken at the same time. I was surprised to see them, especially when I converted the drawing and the picture into black and white. The two were very similar. I recalled my visit to the Seto Inland Sea. After sunset, color quickly disappeared and the stars began to shine. Everything turned to shades of grey, and the shadows of the islands became vivid. This prompted me consider whether those we refer to as "colorblind" in our society are the ones who can best appreciate the beauty of such scenery.

4.2 Disqualifying Clauses on Disability

In Japan, various acts have had disqualifying clauses on disability, stating that persons with certain disabilities cannot undertake certain occupations. In 2007, the Citizens' Committee to Eliminate Disqualifying Clauses on Disability issued a booklet, which illustrated how various acts have prevented persons with disability from participating in society.

---

[4]Megumi Murakami suggested this story to me.

Ms. Kumi Goto (later Kumi Hayase) graduated from a pharmaceutical college in 1998 despite having a hearing disability.[5] She studied hard to pass the national examination for pharmacists, but her application for the license was denied in September 1999.

This case was the first of its kind in which a person with a hearing disability applied for the license while declaring that she did indeed have a hearing disability. The Ministry of Health, Labour, and Welfare explained that the disqualifying clause in the Pharmacists Law was applied to her case.

As a student, Ms. Goto was good at mathematics and sciences, and was interested in an occupation that made full use of her ability. Her mother was also a pharmacist, which would have influenced her as well. Although she was aware of the disqualifying clause, she considered that a challenge to the law would see the clause repealed.

It is true that for Ms. Goto to become a pharmacist, she needs to communicate with co-workers and patients, most likely by means of writing instead of talking. However, it is logical to assume that hospitals and pharmacies must also be capable of treating persons with hearing disabilities. It is likely that these health facilities will be more effective in treating persons with a hearing disability if they employ professionals with the same disability. In turn, as more persons with disabilities work as professionals, the easier it is for others with disabilities to go out into society. In 2000, Ms. Goto worked for a large pharmaceutical company and analyzed data for medicines sold in the market. However, despite having no complaint with her company, she still wished to work with patients.

Ms. Goto collected more than 2 million signatures to file a petition to abolish the unnecessary clause. The government removed the disqualifying clause for her case in 2001 and she became the first certified pharmacist with a hearing disability. In 2017, Ms. Goto was working at Showa University Hospital as a pharmacist and was in charge of patients with hearing disabilities.

Ms. Fusae Kurihara is a nurse with a hearing disability and is a member of our research team. She also faced a disqualifying clause for nurses because of her hearing disability. Fortunately, the clause was abolished in the year that she took the examination, and she was able to complete her studies and started working for a hospital.

One day, Ms. Kurihara misinterpreted the needs of a patient with terminal cancer. When the patient asked her the reason for her mistake, she told the patient about her hearing disability. The patient then reassured her, "It is a good thing that a person with a disability can still work as a nurse. You are a very precious person who can see things from the same viewpoint as patients," and the patient became emotional at this point. This occurred only a few days before the death of the patient. This episode demonstrates how medical staff with a disability tend to show more empathy and understanding than those without a disability, and the patients feel more comfortable talking to them. This episode also reminds us that automatic disqualification of competent health workers based on a disability makes no sense. Ms. Kurihara

---

[5]This description is, among others, based on the article of Asahi Shimbun on Jan. 28, 2000.

joined our team, Research on Economy And Disability (READ), to research medical professionals with disabilities, and then returned to a hospital to work as a nurse.

Thanks to the efforts of people like Ms. Kumi Hayase and the Citizens' Committee to Eliminate Disqualifying Clauses on Disability, many disqualifying clauses have been abolished. However, some similar clauses remain. In 2018, a man in his 30s sued the Japanese Government after he was forced to resign as a security guard because of a disqualifying clause on those who use the adult guardianship system. He claimed that this action was unconstitutional (*Mainichi Shimbun*, 11 January 2018).

According to the petition, the man had a mild intellectual disability. For the position in question, he had been employed as a security guard by a security company in 2014, but he had also worked as a guard for the previous 10 years. An issue arose when one of his relatives used his identity to secure finance for a new car. At this point, the man decided to use the adult guardianship system, and a guardian was appointed on his behalf. Unfortunately, this move is what caused his disqualification from employment. The man described his feelings at the time, "Something must be wrong. I had to resign from my job without any reason. I want to return to my job as a security guard."

A spokesperson for the security company that hired him also expressed their frustration: "He is a very dedicated and helpful employee. We asked him to resign only because it is designated by law, but we did not want to fire him."

The government has reviewed the relevant legislation and now states that there is no reason to keep the disqualifying clause. The government plans to submit an act to the congress that will abolish all disqualifying clauses related to the adult guardianship system.

Chapter 5
Creating the Market for All

5.1 The City of Eyeglasses

In December, 2017, the author visited the northern part of Fukui Prefecture, which is the dominant region for the domestic production of glasses frames. More than 95% of glasses frames are produced here. The purpose of the visit was to interview the President of Masunaga Optical Mfg. Co., Ltd., which has been in operation for more than 100 years.

Fukui is on the northern sider of Honshu and is part of the coastal area that faces the Sea of Japan. The region is economically underdeveloped compared with the Pacific coastal areas. In winter, it is covered with snow, which typically lays more than one meter deep. After widespread economic and social changes brought about by the Meiji restoration, the Fukui area continued to lag behind other more developed regions. More than 100 years ago, towards the end of the Meiji era, Gozaemon Masunaga, a wealthy farmer in Shono, Fukui, welcomed back his younger brother Kohachi on his return from Osaka, the second largest city of Japan. Kohachi possessed a knowledge of fashion and persuaded Gozaemon to start a new business to manufacture eye glasses. Although Gozaemon himself had never seen glasses before, he had always considered ways to improve the lives of the villagers, and agreed to start the business.

The next step for the two brothers to take was to find good craftsmen. They sought the help of Suekichi Masunaga, who initially declined their offer. Suekichi was a carpenter building shrines and had no experience in small-scale craftsmanship required to make glasses. Furthermore, like Gozaemon, Suekichi had never seen glasses before and had no idea how they might be useful.

In spite of this response, history was irreversibly changed by Tsune, the daughter of the carpenter.[1] The brothers visited Suekichi on a weekday, but, to their surprise, Tsune was also at home, playing by herself. Kohachi questioned why Tsune was at

---

[1] The story of Tsune is in the novel "*Oshorin*" by Yoko Fujioka, which is essentially based on a true story.

© Springer Nature Singapore Pte Ltd. 2019
A. Matsui, *Economy and Disability*, Economy and Social Inclusion,
https://doi.org/10.1007/978-981-13-7623-8_5

home even though she was a school-age girl, and the school must be open that day. Of course, it was common around that time that poor families do not send their children to school. However, the carpenter's family did not seem to be so poor as to take that option. When Kohachi questioned, Suekichi explained that Tsune was asked to leave school since she could not keep up with classwork. He said, "She could not even write down what the teacher wrote on the blackboard."

Watching how Tsune was playing, Kohachi could not believe that she could not cope with elementary school. There must have been a different reason, Kohachi thought. He then took out some sample glasses from his bag and asked her to put them on. Tsune then screamed, "The light is too strong..." She closed her eyes with surprise. She then opened her eyes again and said, "Father, your face looks different."

It was clear to Kohachi that Tsune was a clever girl, but had a poor eyesight and could not read what the teacher wrote on the blackboard. In Fukui in the early 1900s, there was neither the concept of poor eyesight as long as one can send a daily life nor the observation that wearing glasses would correct eyesight. Therefore, anyone that could not see characters on the blackboard was considered the same as one who could not follow classwork.

Suekichi was deeply moved and decided to cooperate with Gozaemon. That was the moment when the glasses industry was born in Fukui.

After Suekichi agreed to make glasses, Gozaemon recruited some other young people and started making glasses frames. He invited a craftsman from Osaka, and once the craftsmen at Masunaga mastered a range of techniques, he turned to another craftsman with higher-level skills. After the Masunaga craftsmen reached a suitable skill level, Gozaemon nominated the first generation of *Oyakata* (supervisors) and gave them their own workshops. Competition then started between the workshops, and within a few years, one of the workshops won a gold medal in a nation-wide exhibition.

Since that time, the Fukui region has emerged as the dominant area for the manufacture of glasses frames. Fukui workshops initiated the use of nose pads to improve the fit of frames on Japanese faces and they also succeeded in making titan frames. The initial business strategy was to become an original equipment manufacturer (OEM) to make frames for other well-known firms. This approach is a convenient way to produce goods because the OEM does not have to create its own brand, which takes time and patience as well as cost. However, without its own brand, the dependency of OEMs on other well-known firms can be a problem when a firm no longer requires the services of an OEM.

Indeed, such issues emerged in the late 1990s and into the 2000s. In the early 1990s, the domestic retail market for glasses was about 600 billion yen (6 billion USD), but by 2010 the market had fallen to 400 billion yen (4 billion USD). This contraction was mainly caused by competition, especially with products from China. This situation was the same for Masunaga. In 2009, when establishing its new brand, Masunaga revived classical frames, which became very popular in France. In 2013, Masunaga G.M.S. Limited won the Silmo d'Or Special Jury Prize, which is one of the most prestigious exhibition prizes in the world. Currently, more than 50% of Masunaga sales are in the export market.

In spite of its success, Masunaga Optical has maintained its tradition. The company's mission statement since its establishment has been, "We manufacture excellent eyeglasses. We want to make a profit if we can, but we don't hesitate to take a loss. It is always in our thoughts to manufacture excellent eyeglasses."[2] These words are still the basis of the code of conduct for employees after more than 100 years. Thanks to the manufacturers of eyeglasses such as Masunaga Optical, those who have poor eyesight are no longer considered as persons with a disability.

5.2 Atelier Incurve

Atelier Incurve is situated in the outer suburbs of Osaka, the second largest city of Japan. Osaka is a big city, but a 30-min ride from the city center will take you to open suburban areas where rice fields and other crops are visible among the houses. Atelier Incurve is located in such a setting. Unlike the surroundings, the Atelier Incurve building looks like an expression of modern art. Inside the building, an abundance of light and space in no way suggests this is a welfare facility. As I entered the building, a man with a familiar smiling face greeted me. The man was Mr. Hiroshi Imanaka. He is the manager of the facility, but looks more like an art director. Indeed, he is the founder of this atelier and the designer of its space. As soon as I entered his office, I noticed a particular painted picture.

Yosakoi Matsuri by Takeshi Sakamoto

---

[2]This quote is taken from the home page of Masunaga Optical.

The painting depicted seven figures dancing in a festival called *Yosakoi Matsuri*, which is one of the major festivals in Japan. All the figures were different in shape and face. A figure in the center of the painting had the round face of a tiger but with oranges for ears; and yet it looks as if a rice cake is dancing. A figure in the right bottom corner looked like a girl, but also like a deer. Mr. Imanaka explained, "The painting is a work of Mr. Takeshi Sakamoto. He is a rising star here."

The purpose of my visit was to interview Mr. Imanaka about an art market that had been created by the artists at Atelier Incurve and other studios. Atelier Incurve was founded in 2002 as a government-funded initiative to support the creative efforts of artists with intellectual disabilities. Although some people consider that artwork created by artists with intellectual disabilities is done for welfare purposes, such a perception probably deprives the artists of the opportunities of participating in the market. The artwork made by the artists at Atelier Incurve certainly deserves appraisal in the market. The market is not a jungle where the strong prey upon the weak, but rather a place that empowers persons with disabilities. Such empowerment is the goal of Atelier Incurve.

After the interview, our discussion returned to Mr. Sakamoto's picture and I asked to see some other examples of his work. All of the figures in his artwork were humanized animals that had round eyes that stared out from the picture. I also appreciated his expert use of color, but I felt a connection with the earlier painting. I liked the background color, which was brownish orange, and I concluded that it would look good on a wall in my house.

At Atelier Incurve, the assistants who supported the artists were not originally trained as social workers, but rather as artists or art curators, although most had since obtained a certificate in social work. The number of artists at the time of my visit was 26. All of them were registered as persons with intellectual disability. In spite of this, when I listened to them explaining their art, it was not obvious that they had intellectual disabilities.

On another visit to Atelier Incurve, I met another remarkable artist. "This is Mr. Terao," Mr. Imanaka explained. I felt absorbed into the picture he was drawing. The picture was of a massive steel-framed construction. Mr. Katsuhiro Terao is amazingly productive. He explained to me, "I'm working on this, while drying another one." Mr. Terao sometimes completes a 2 m × 2 m painting in about 2 weeks. In the market, such paintings sell for about 4 million yen, or about 40,000 USD. Mr. Terao receives the sale proceeds after deductions for the cost of the sale.

Mr. Terao had worked as a welder at his father's factory and all of his works are related to his earlier vocation. According to Mr. Terao, his artworks are blueprints rather than pictures. "Mr. Terao has a good memory of his father's factory," Mr. Imanaka explained.

On another occasion, a lecture was given by Mr. Terao at the Kanazawa College of Art, although in some ways it was more like a performance. The floor was totally silent because of the concentration of Mr. Terao, but many questions followed in the question-and-answer session. One female student asked if he ever became bored with drawing because he seemed to be drawing the same thing again and again. Mr. Terao replied, "I don't get bored as I like what I do." The student became emotional at this point, and some show of emotion spread among the audience. She student replied, "I have liked drawing from my childhood... That is why I came here... But now, I draw to please my professors or to have it praised by others... I admire Mr. Terao who can say he draws because he likes it."

Mr. Imanaka himself studied design and worked for a large design company. He explained, "Mr. Terao and others do not change their minds. They stick to one thing." Mr. Imanaka had often thought about the concept of originality in art and then he encountered Mr. Terao and his artwork. He thought, "I cannot create such original art." At the same time, Mr. Imanaka was surprised to find that Mr. Terao's wage was below the minimum legal wage because he had an intellectual disability.

Original talent is not cultivated only by education and competition. "Education is an obstacle," Mr. Imanaka firmly explained. Those who have been educated in a stereotypical manner—who have been told to follow the directions of persons without intellectual disability—cannot work on art even if they are told to draw freely. If they are given directions, their originality withers away as they wait for further directions. Also, if only competition is in place, then all the buds of creativity will be pulled off.

While securing a basic stipend for daily life by way of welfare programs, the artists whose talents come into bloom are able to compete in the market. Differences in terms of sales and income can then emerge, and people who put priority on equality in outcome often criticize it. However, should one be restricted to the average life of the group that he/she belongs to once he/she is labeled as a person with intellectual disability? Few people complain about the high earnings of star baseball players, and we explain it away as a reward for their talent. The criticism that some people with intellectual disabilities earn money because of their talent is based on envy. "He (Mr. Terao) earns more than me," says Mr. Imanaka, laughing.

One may argue that persons with disability live on welfare, but welfare is covered by tax just as roads, security, and other public goods are funded by tax. Thus, not only those with disabilities but all of us depend on government funding to some degree. If these artists can draw pictures that are sold abroad, they contribute to the economy and society in a very real way.

Indeed, Mr. Terao is a global figure even though he never goes abroad. His pictures sell for 40,000 USD in the US art market. While at Artfair Osaka, I was thinking of buying a small picture for 50,000 yen when a French art dealer stopped by the exhibition room, looked around, and immediately bought the largest of Mr. Terao's pictures. A true global figure has the ability to attract people from around the world.

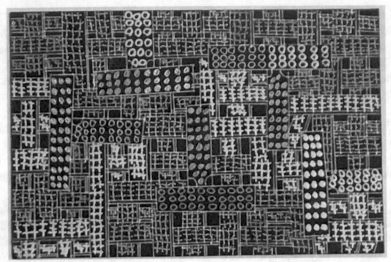

'Bolts' by Katsuhiro Terao.

   Mr. Terao was at Artfair Osaka, but he did not seem overly concerned about the art sales. His attitude was reminiscent of Diogenes, a famous Greek philosopher, who never lost sight of ordinary things. It is reported that Alexander the Great once visited the cave where Diogenes lived. It would have been extremely rare, of course, that the king of kings would go out of his way to make such a visit, rather than issuing a summons for the philosopher to visit the king. Their conversation was also surprising. The great king, standing in front of the cave, asked Diogenes, "Tell me your wish, and I will fulfill it." Diogenes replied, "Would you step aside so that the sunlight comes into my cave?" Mr. Terao never loses sight of the things that matter, much like Diogenes. He loves freedom and hates interference.

After leaving Osaka, I visited *Yamanami Kobo*, another art studio in Koga-City, Shiga Prefecture, also known for the village of Ninja. The studio was located in a mountain area and was covered with snow. This studio accommodates persons with intellectual and mental disabilities, but at *Yamanami Kobo* they are artists.

   In the studio, the artists were working unsupervised. I observed Mr. Inoue drawing minute patterns on a huge canvas, while, nearby, Mr. Yoshikawa was making small holes in porcelain clay using a chopstick. Then Ms. Ohara, who was buried in a sofa, suddenly got up and started making small figures of monks, working quickly. I was told that she spent only one hour each day on making the figures, but the stockroom was full of her artwork. I was attracted to a picture by Mr. Kurita. I was told that his mental state has ups and downs and this can be reflected in his artwork: the pictures drawn when he is calm are totally different from those drawn when he is uneasy. I shared a picture drawn by a "calm Kurita," and one of my friends immediately responded, "I'll take it!"

The market is a place where a person meets goods and another person. Artworks also need a market, and without one we would miss opportunities to experience and perhaps buy wonderful artwork. Through the action of the market, people who were previously unknown to each other are suddenly linked. This fact itself is part of the power of the market.

In the course of this work, I was able to meet Ms. Shino Sugimoto, a curator and president of Foster Co, who was working to create such a market. After holding an exhibition of artworks from Atelier Incurve, the success of that exhibition prompted Ms. Sugimoto to have another, much bigger, exhibition in 2017. The exhibition was held at GYRE in Omotesando, Tokyo, which is the center of action in Tokyo. I visited the exhibition a couple of weeks after its opening and was surprised with what I saw. More than half of the artworks had already been sold. Moreover, a high proportion of the bigger and more expensive artworks had "sold" signs attached. These larger artworks were priced in the order of hundreds of thousands. Ms. Sugimoto told me that the prices of the artworks at the exhibition were comparable to those for artwork in the standard art market. Among those sold was a huge picture by an "uneasy Kurita".

There is no official way to become a participant of the market. Through the market, you can be a patron of an artist, you can simply love the art, or you can be an investor. Your preference and behavior should be determined by yourself. That is the nice thing about the market. Like the artists that create their artworks as they see fit, we can participate in the art market in our own way.

Let me return to the painting of Mr. Sakamoto (*Yosakoi Matsuri*). They said they were going to send it to an artfair in New York as one of the headline artworks. With only a short time before the artwork left Japan, I almost bought it on the spot. After confirming the availability of the item and a quick check of my hanging space, I quickly replied to Mr. Imanaka: "I'll take it!"

Each transaction such as this one creates a market for the artist. The market can be a physical place like an art fair, but every occasion where a seller meets a buyer constitutes a market. I met the picture by Mr. Sakamoto and was fascinated by the round eyes.

The market can be hard, but honest. It does not flatter you. Those who emphasize the importance of supporting persons with disability will not pay the market price for artwork unless they consider the art to be worthy of the price. The market may induce uneven outcomes, but, at the same time, it creates a dream. For one artist, who used to work at a facility with simplistic work before coming to Atelier Incurve, his mother explained, "If my child sparkles like a one-time firework once in his lifetime, I'll be content with it. That is so even if he disappears afterwards."

It was pleasing to see the multiple effects of Atelier Incurve, and I felt that the market allowed these individuals to shine as artists rather than as people with disabilities. To me, they were more like stars rather than one-time fireworks.

5.3   Children Attending *Juku*

Tutoring schools, or *juku*'s, constitute a huge education industry in Japan. Many children go to *juku* after school. In 2013, about 4.1 million people, most of whom were children in elementary, mid-high, and high schools, went to *juku*.[3] This number does not include those of home tutors, correspondence studies, and lessons for enrichment and English conversation. The number of school-age children in 2013 year was 13.5 million,[4] which means that about 30% of Japanese school children go to *juku*. Some children study more challenging materials in *juku* than at school, while others study the materials that have already taught at school. The same *juku* often offers both courses to different levels of children.

The number of *juku*'s in 2013 was over 50,000, and the total sales of these schools amounted to 970 billion yen, or about 10 billion USD. Some *juku* operate as nation-wide businesses. Benesse Holdings, the largest of all, had total sales of 444 billion yen in FY 2015/2016. On the other hand, there are more than 29,000 *juku* with four employees or less.

Susumu Murakami is an owner/manager of a *juku*, where he is the only teacher. This *juku* is located in an industrial city. About 100 children, ranging from fifth grade to senior in high school, attend Murakami's *juku*. The abilities of the children vary. The top-level students tend to study for top schools including the University of Tokyo and Kyoto University, while, on the other hand, some are high school dropouts. Because Mr. Murakami is the only teacher, the most important factor in the *juku*'s operation is the personal relationship between Mr. Murakami and the children. Some children quit on the first day, while others eventually go on to college even though they were high school dropouts.

Ichiro (alias) was a junior high student when he joined Mr. Murakami's *juku*. Although, he was considered a troublemaker at school, he regularly attended the *juku*. After dropping out of high school, Ichiro returned to the *juku* and indicated that he wanted to go to a different high school. Given that Ichiro was good at mathematics, Mr. Murakami continued to develop his mathematical ability, and Ichiro gained entry to the new high school. Ichiro also possessed sporting ability and used his physical stature to his advantage to become a star rugby player at the new school. Nonetheless, he continued to attend the *juku*. Mr. Murakami suggested that Ichiro should go to a private college in a nearby city. Although its admission did not require an examination on any subject, it did require an essay based on a book of the student's choice. Mr. Murakami had Ichiro read only one book thoroughly so that he could answer any question. Ichiro then passed the exam and gained entry to the private college. Later, Ichiro visited Mr. Murakami on the coming-of-age day, dressed in his suit. Ichiro graduated from the college, and now works at a small factory.

---

[3]The data of *juku* are based on the report on the *juku* industry of NTT town page: http://www.ntt-tp.co.jp/lab/gyoukai/education/10/; accessed Feb. 12, 2018.

[4]*Gakko Kihon Chosa* FY H25.

Misa (alias) was a junior high student who could not go to school, but she still attended Mr. Murakami's *juku*. As problems emerged for Misa at school, she would report to the office of the school nurse. Eventually, her home room teacher instructed her to go to the classroom and to stop attending the nurse office. At this point, Misa stopped going to school. The teacher encouraged her to go back to school and made the effort to call Misa every day. However, this situation only added to the pressure that Misa felt. Although Misa had trouble managing pressure, she was still well behaved and possessed a good work ethic. When attending school, she typically woke at 5.30 a.m. to properly prepare for the day. Nevertheless, Misa continued to attend Mr. Murakami's *juku* and passed an entrance examination for a private high school. However, once again, Misa found that she could not go to school at all. After one year, she finally received a holdback notice from the school. Mr. Murakami then suggested that she step back from the situation, perhaps take a holiday, or do something else to relax. Misa did not agree, saying that she should study hard in a correspondence course and go on to college in three years. She now works in a part-time job in the morning and studies at Mr. Murakami's *juku* to work on a correspondence course.

Juku can be different things to different people, but they can be a refuge for children who have no other place to escape. At home, parents have all the power, and similarly at school, teachers have all the power. When a child clashes with parents or teachers, it is difficult for the child to perform well. Many different issues might be blamed on the behavior of the student, rightly or wrongly, and the situation can deteriorate quickly. In such cases, the *juku* can offer an effective alternative. Mr. Murakami explained that he often reassures students: "You are not bad. Your parents make mistakes. Sometimes, your teachers say the wrong thing too."

While we may regard school as a public place (including private schools as well), *juku* may be considered as a private place–in other words, a version of the market place. No matter how good a school may be, there will be some fraction of students who cannot cope with school life. *Juku* creates places for such students. It would be ideal if conventional school met the needs of every student, but this is not the case in reality. Perhaps we should accept this system–there will always be those that have trouble with school–maybe an education system with no school dropouts would be a sign of a deeper problem within the system.

From a personal perspective, I do not argue that *juku* is better than school. As a student, I disliked *juku* and never attended on a regular basis. The main point we can take from the existence of *juku* is that it is a product of the market, where the market addresses the deficiencies of the school system. *Juku* is entirely optional–if you do not like it, you do not have to go. On the other hand, school is not optional, and you cannot stop going to school without incurring some cost.

When asked why he became a *juku* manager, Mr. Murakami replied, "There must be a variety of reasons. Most people do not choose *juku* lecturer as their vocation from the beginning. In my case, I did not want to follow the role of a *juku* manager, so I became independent." He recalled that he did not cope well at school: "There were some respectful teachers, but I felt isolated at school. These respectful teachers criticized the school system. One of them even told me not to become a teacher." He went on, "I do not have any aspiration of making Japan or Japanese education better. I like teaching, I have better teaching skills than some, and I can save some children by way of my skill. I do not try to accommodate all. It would be nice if 51% of the students like my way of teaching. I cannot save thousands of students, but I can save hundreds of them." I saw his pride in his words.

Part II
Game Theory and Disability

Chapter 6
Game Theory

This chapter introduces game theory in a non-mathematical fashion, although some simple math is used, and disability-related issues are used as examples. The main purpose of this chapter is to explain the concept of game theory, especially for readers who are not familiar with it.

6.1 Introduction

As the world economy is now integrated more than ever, and no single country is itself a dominant force, issues associated with interaction between different economic and social systems begin to appear. The economic success of post-war Japan has contributed to this curiosity. After traditional economics, which often assumes a particular form of institution as given, could not satisfactorily explain these phenomena, new approaches have emerged, using game theory, contract theory, and information theory to better understand the rationality hidden behind seemingly special arrangements of Japanese firms. This line of research is discussed in a survey by Aoki (1994), who emphasizes the importance of analyzing the Japanese economy as a system rather than by studying several factors one by one. Because of the strategic complementarity exhibited between various institutions, different systems may be sustained as equilibria in different regions even if the "physical environments" are similar. This observation gives rise to further questions such as: if the present "physical environments" are similar across developed countries, what causes the observed differences in the systems? Or more simply, how did different systems emerge in the first place?

A new line of research has been developed in game theory known as evolutionary game theory. This theory has been developed by those who cast doubt on the idea of perfectly rational agents, which has been widely assumed in the study of human economic behavior. Since Nash equilibrium was proposed by Nash (1951), its refined

© Springer Nature Singapore Pte Ltd. 2019
A. Matsui, *Economy and Disability*, Economy and Social Inclusion,
https://doi.org/10.1007/978-981-13-7623-8_6

concepts have been proposed by many researchers to capture the behavior of perfectly rational players. The more refinements of Nash equilibrium were proposed, however, the clearer it became that the study of perfect rationality alone would not lead us to an understanding of human behavior. In addition, some people began to cast doubt on the existence of the very goal of this work, i.e., the existence of a single notion of "perfect rationality." Evolutionary game theory does not necessarily require agents to be "rational" and puts more emphasis on what has been established in a society. Although evolutionary game theory still remains at the stage of theoretical development without much application, the above feature indicates its potential usefulness in the study of different social systems.

We will introduce evolutionary game theory as a tool for analyzing issues on the diversity, interaction, and evolution of social systems and disability-related issues, discussing some of the results obtained in the field. For this purpose, the present chapter takes a course of introducing only the papers that are relevant for our discussion. Those who are interested in evolutionary game theory itself should refer to other papers and books, including Crawford (1991), Hammerstein and Selten (1994), Hofbauer and Sigmund (1998), Mailath (1992, 1993, 1995), Sandholm (2010), van Damme (1987, 1994), Vega-Redondo (1996), and Weibull (1997).

6.2   Social Norms and Rational Behavior

Evolutionary game theory is important in the analysis of social systems. The argument is twofold; one is conceptual and the other practical. We first examine the connection between evolutionary game theory and the analysis of social systems from a conceptual viewpoint.

Issues on culture and social norms have been discussed extensively in the social sciences. Different disciplines view the issues from different perspectives. For example, social norms are often regarded by sociologists as orientations by which individuals are guided. An orientation becomes a social norm when it is shared by most members in a society. However, economists tend to view a social norm as an equilibrium. Incentive problems are the main issue here. Elster (1989) writes about this discrepancy:

> One of the most persisting cleavages in the social sciences is the opposition between two lines of thought conveniently associated with Adam Smith and Emile Durkheim, between homo economicus and homo sociologicus. Of these, the former is supposed to be guided by instrumental rationality, while the behaviour of the latter is dictated by social norms. The former is "pulled" by the prospect of future rewards, whereas the latter is "pushed" from behind by quasi-inertial forces.

Evolutionary game theory sheds new light on this dichotomy. A typical model in this field assumes inertia, whereby it takes time to adjust one's behavior. However, agents are not blind followers of customary behavior. They adjust their behavior as the environment changes, and the environment is in turn affected by human behavior. Agents sometimes take a strategy that maximizes the discounted sum of their expected future payoff. In other words, agents are pulled by the prospect of future rewards, and, at the same time, pushed from behind by quasi-inertial forces.

Let us elaborate on the relation between this dichotomy and evolutionary game theory. Elster (1989) "define(s) social norms by the feature that they are not outcome-oriented. The simplest social norms are of the type 'Do X', or 'Don't do X'. (p. 98)" In the context of evolutionary game theory, social norms defined this way are viewed as the major force that makes individuals keep their past mode of behavior.

On the other hand, changes in behavior patterns are caused by two forces. The first is outcome-oriented behavior, and the second is random shock. A simple form of outcome-oriented behavior is expressed as "Do X if X leads to Y." Outcome-oriented behavior includes rational behavior as an important example; however, the former by no means implies the latter, however. Another example is adaptive behavior by boundedly rational agents. The other force of change, random shock, is often referred to as mutation. Sociologists often describe those who induce random shock as deviants rather than mutants. In sociology, deviants are those who do not conform to the social norms of their society. In spite of the negative connotation associated with the term "deviants," whether deviants are harmful or useful depends on the situation, one's point of view, and so forth.

Evolutionary game theory is useful, from practical and conceptual points of view, in studying the diversity, interaction, and evolution of social systems. To see this, consider the following features of social norms. Social customs are often formed because coordination of behavior is necessary. Without social norms, a society will experience social disorder almost by definition. Every society has its own set of social norms. This is because, in many situations, coordination on some behavior is more important than the intrinsic value of behavior.

In such a situation, multiple equilibria arise, and equilibrium selection becomes an issue.[1] It is in precisely this type of situation that evolutionary game theory has an edge over traditional game theory based only on rational behavior since history is an explicit factor in evolutionary game theory.

6.3 Games in Strategic Form

6.3.1 Games in Strategic Form and Zero-Sum Games

A game form should contain three components:

1. Players;
2. The options available to them (strategies);
3. The consequences of their behavior (payoffs).

---

[1]The purpose of equilibrium selection is not necessarily to select a single outcome no matter what the history may be. Rather, its purpose is to shed light on the mechanism by which an equilibrium is selected.

Let us consider a simple game of matching pennies considered by von Neumann and Morgenstern (1947) as well. In this game, there are two players, A and B. Each player has a coin and presents either head or tail at the same time. If the two coins match (either two heads or two tails), then A wins one point (and B loses one). If the two coins do not match (one head and one tail), then B wins one point (and A loses one).

This is a well-formulated game that has the three necessary components as follows:

1. Players: A and B;
2. Strategies: Heads and Tails;
3. The consequences of their behavior: Player A wins one point if the pennies match, while Player B wins one point if the pennies do not match.

This situation is better described in a tabular form. Table 6.1 captures all the aspects of the matching pennies. Player A is often referred to as the row player, who decides which row, Heads or Tails, to take. We call Heads and Tails (pure) strategies. Player B is the column player, who decides which column to take. The consequences are expressed in terms of payoffs. For example, the entry that corresponds to (Heads, Heads) leads to the outcome $(1, -1)$, meaning that A obtains 1, and B obtains -1. Other entries are interpreted in the same manner. It is a convention that in each entry, the number on the left is the payoff to the row player, and the number on the right is the payoff to the column player.

The solution to this problem is intuitive: each player takes Head and Tail with the equal probability of $1/2$ (we call it a mixed strategy). To see this, let us consider first what happens if, say, Player A always takes Heads. If this is known to Player B, then Player B will take Tails to obtain 1. But, if that is expected, Player A should take Tails. This type of reasoning forms a cycle.

A similar cyclical argument can be made if at least one of the two takes Heads and Tails with uneven probabilities. If, for example, Player A takes Heads more often than Tails, then Player B is better off by taking Tails, to which Player A should respond by taking Tails.

If the both players take Heads and Tails with equal probability, then the other player cannot take advantage of the unevenness of the opponent. Therefore, this pair of mixed strategies is considered as a solution to this game.

The game of matching pennies is classified as a two-person zero-sum game. The game is called a zero-sum game since the sum of the two players' payoffs in each entry is always zero. von Neumann and Morgenstern (1947) fully analyzed the class

Table 6.1 Matching pennies

A	B	
	Heads	Tails
Heads	1, −1	−1, 1
Tails	−1, 1	1, −1

of two-person zero-sum games. The next major development was the celebrated solution concept of Nash, now called Nash equilibrium.

6.3.2 Nash Equilibrium

Since Nash (1951) proposed a solution concept, now known as Nash equilibrium, its interpretation and refinement have been discussed extensively. A strategy profile is called a *Nash equilibrium* if no player has a strict incentive to deviate from the designated strategy given that other players will follow their prescribed strategies. Formally, consider a normal form game,

$$G = \langle I, (A_i)_{i \in I}, (u_i)_{i \in I} \rangle,$$

where I is the finite set of players, A_i $(i \in I)$ is the finite action set of player i, and $u_i : A \to \mathbb{R}$ is the utility function of player i with $A \equiv \times_{i \in I} A_i$. We then allow the set $\mathcal{F}_i$ of mixed strategies of player $i \in I$, which are probability distributions over A_i. The utility function u_i is extended to the expected utility function Π_i. Then $\sigma \in \mathcal{F}$ is a Nash equilibrium if for all $i \in I$, and all $\rho_i \in \mathcal{F}_i$,

$$\Pi_i(\sigma) \geq \Pi_i(\sigma_{-i}; \rho_i)$$

where $(\sigma_{-i}; \rho_i)$ is a strategy profile that is obtained from σ by replacing σ_i with ρ_i.

Among the various interpretations of Nash equilibrium (see, e.g., Aumann 1987; Binmore 1987, 1988; Kaneko 1987), Nash seems to adopt the rationality interpretation, which corresponds to the eductive processes in Binmore's terminology. In this interpretation, the players are sufficiently rational to be able to analyze a one-shot game to be played (if it is a repeated game, the repetition occurs only once). In his paper, Nash identified a class of games in which the equilibrium analysis is more valid than the rest (Nash solvability) and proposed its refinement (symmetric equilibrium). Also, both Nash solvability and symmetric Nash equilibrium are motivated by the rationality interpretation. Roughly speaking, a game is said to be Nash solvable if each player is indifferent between his actions used in some Nash equilibria as long as other players take such equilibrium actions. If a game is Nash solvable, then the equilibrium analysis is consistent with the rationality interpretation. No selection problem arises, and all players can play optimally if Nash's theory is believed, and if such a belief is common knowledge.

Symmetric Nash equilibrium is also motivated by the rationality interpretation. A symmetric Nash equilibrium is defined as a Nash equilibrium such that the same probability is assigned to any symmetric pair of actions, a pair whose permutation gives the same game as the original game. This notion is based on the observation that any identical actions, as well as any identical players, should be treated equally. Symmetric Nash equilibria always exist. The reason that this refinement is better viewed as one based on the rationality interpretation is seen in the following example

Table 6.2 Pure coordination game

	L	R
T	1, 1	0, 0
B	0, 0	1, 1

(see Table 6.2). In this game, known as a pure coordination game, every pair of actions is symmetric. Thus, the probabilities attached to T, B, L, and R must be the same. The only equilibrium that satisfies this requirement is the mixed strategy Nash equilibrium $(\frac{1}{2}[T] + \frac{1}{2}[B], \frac{1}{2}[L] + \frac{1}{2}[R])$, i.e., the one where each action is taken with the equal probability of a half. Symmetric Nash equilibrium is best understood in a situation where players play this game for the first time. Players who have not played this game before have no reason to distinguish between, say, T and B. As a result, they end up playing the mixed strategy Nash equilibrium. However, this equilibrium is least likely to sustain itself if the game is repeated in a large population since the behavior pattern tends to move away from it if the strategy distribution fluctuates a little in either direction, L or R. Indeed, evolutionarily stable strategies (Maynard Smith and Price 1973) are $([T], [L])$ and $([B], [R])$. While often more sophisticated, a subsequent series of refinements of Nash equilibrium were based on the rationality interpretation as well.

6.4   Evolutionarily Stable Strategies

When one discusses equilibrium refinement, he/she sets forth some requirements that each equilibrium should satisfy. In such an attempt, some adjustment process is implicitly or explicitly considered. In equilibrium theory, it is unclear what kind of force induces an equilibrium. Binmore (1987, 1988), among others, consider the problem of classifying adjustment processes into two categories, eductive processes and evolutive processes. According to Binmore, eductive processes occur in notional time and require sufficient knowledge and ability of players to analyze a game prior to play, which is repeated only once (if it is a repeated game, then repetition occurs only once). Backward induction operated in a player's mind is an example of an eductive process. On the other hand, evolutive processes occur in real time. A certain situation is repeated many times, and players are not necessarily sophisticated enough to analyze the game. Rather, they adjust their behavior gradually according to some rule. The dynamical processes considered in evolutionary game theory are close to the evolutive process, although there is no clear definition of "evolutionary game theory."

We confine our attention to two-person random matching models. This means that we consider a society consisting of many (often anonymous) agents who are randomly matched to form pairs and play two-person games. We also assume that agents are anonymous, and that any matched agents will never be matched in the future. When we consider such a society, a description of the game is not sufficient to specify the situation. There are two classes of matching situations to be distinguished. The first is the class of situations in which two players are chosen from the same pool of individuals. The second class consists of situations in which two different types of individuals, such as male and female, are matched. We call the former a matching of the form M^1 and the latter of the form M^2.

Once again, let $\mathcal{F}_i$ ($i = 1$ for M^1; $i = 1, 2$ for M^2) be the set of mixed strategies of a type i player. We identify this with the set of strategy distributions of type i players. We write $\mathcal{F} \equiv \mathcal{F}_1$ for M^1, and $\mathcal{F} \equiv \mathcal{F}_1 \times \mathcal{F}_2$ for M^2. Given a strategy distribution $f \in \mathcal{F}$ and a type i player's mixed strategy g_i, let $\Pi(g_i; f)$ be the expected payoff to the type i player taking g_i and facing f. We often remove the subscript "i" when we consider M^1. We bear in mind the interpretation in which each type consists of a sufficiently large number of agents who are anonymous and who are matched randomly at every instant. Without this interpretation, the following arguments will have little validity.

Maynard Smith and Price (1973) proposed evolutionarily stable strategies to capture the stability of biological adjustment processes. Consider a matching of type M^1. The payoff matrix is often called a fitness matrix in evolutionary game theory: the first number of each entry in a payoff matrix is the fitness level of an agent who takes the corresponding row and faces an opponent taking the corresponding column. A strategy distribution $\sigma \in \mathcal{F}$ is an *evolutionarily stable strategy* (*ESS*) if for all $\tau \neq \sigma$,

$$\Pi(\sigma; \sigma) \geq \Pi(\sigma; \tau), \tag{6.1}$$

and

$$\Pi(\sigma; \sigma) = \Pi(\sigma; \tau) \implies \Pi(\tau; \sigma) > \Pi(\tau; \tau). \tag{6.2}$$

In this definition, $\Pi(\sigma; \tau)$ is the payoff to the agent taking τ against the population taking σ.

Condition (6.1) requires that an ESS is a Nash equilibrium, which is often called a symmetric Nash equilibrium in evolutionary game theory, different from the symmetric Nash equilibrium of Nash (1951) discussed above. Conditions (6.1) and (6.2) together imply that if a small fraction of mutants τ invade the original population σ, the original population is strictly better off than the mutants in a mixture of σ and τ.

The concept is used to analyze a match such as the one given by Table 6.3 (Maynard Smith and Price 1973). If we replace the strict inequality in (6.2) with a weak inequality, then we obtain the definition of weak ESS's (Thomas 1985).[2] In this

Table 6.3 Hawk-Dove game

	D	H
D	2, 2	1, 3
H	3, 1	0, 0

---

[2] In the sequel, we use the same action set for the row and the column players in M^1, but use different sets in M^2.

Table 6.4 A game with
redundant strategies

	L	C	R
L	2, 2	2, 2	0, 0
C	2, 2	2, 2	0, 0
C	0, 0	0, 0	0, 0

game, the action H yields a very high payoff if the opponent concedes to play D.
However, it leads to a disaster if the opponent also plays aggressively. D is a "safe"
strategy, but it is exploited by the opponent who is playing H. In such a game neither
H or D can be stable by itself. In the population consisting of H, agents harm each
other, and playing D will be paid off in such a situation. While in the population
consisting only of D, an opportunist plays H and exploits other agents. The unique
ESS in this game is $\frac{1}{2}[H] + \frac{1}{2}[D]$.

ESS's do not necessarily exist. Consider a match M^1 with its game given by
Table 6.4. This game has two identical actions, L and C. Since they are identical,
every strategy distribution is invaded either by L or C. Even worse, if we eliminate
a redundant strategy C, then there is a unique ESS [L]. This is not a desirable result
since a redundant strategy should not change results in a substantial way. One way
to cope with such a drawback is to consider a set-valued concept. Thomas (1985)
introduces the concept of an evolutionarily stable set. A closed set $F^* \subset \mathcal{F}$ of Nash
equilibria, i.e., strategies satisfying (6.1), is an *evolutionarily stable set (ES set)* if
for all $\sigma \in F^*$, there exists a neighborhood $U(\sigma)$ such that for all $\tau \in U(\sigma)$,

$$\Pi(\sigma; \sigma) = \Pi(\sigma; \tau) \Rightarrow \Pi(\tau; \sigma) \geq \Pi(\tau; \tau),$$

where the equality of the consequent holds only if $\tau \in F^*$. That is, Thomas's ES set
is a closed set of weak ESS's that is robust against mutation outside the set in the
sense of (6.2).

By making the concept set-valued, we avoid the problem associated with redundant strategies, but we still cannot assure existence. See Table 6.5 (van Damme 1987).
In this game, a unique Nash equilibrium is $\sigma^* = \frac{1}{3}[L] + \frac{1}{3}[C] + \frac{1}{3}[R]$. Therefore,
the only candidate for an ES set is $\{\sigma^*\}$ as a singleton. However, it is not an ESS
since strategy [L] fares better than σ^* in a population with a large fraction of σ^* and
a small fraction of [L]. It seems that we must abandon Nash equilibrium if we are to
assure existence.

6.5   Deterministic Dynamics

ESS is a static notion, even though its motivation is to capture some dynamic stability.
For ESS to be valid, it must be related to some dynamical process. It is known that a
strategy profile is an ESS if and only if it is an asymptotically stable point of a simple
dynamic in a population in which all members take the same strategy (monomorphic

population). Consider $\sigma \in \mathcal{F}$ and an arbitrarily chosen mutant τ. Let $\mu(t) \in [0, 1]$ be the fraction of individuals taking σ at time t. Suppose that the rest are taking τ. Then the strategy distribution at time t is given by $\sigma^\mu(t) = \mu(t)\sigma + (1 - \mu(t))\tau$. Now consider the dynamical process given by

$$\frac{d\mu}{dt} = \Pi(\sigma^\mu(t); \sigma) - \Pi(\sigma^\mu(t); \sigma^\mu(t)), \quad \mu(0) = \mu_0 : \text{ given}, \qquad (6.3)$$

where the process stops when it reaches an end point. In this process, the individuals that take σ increase their share if and only if their payoff exceeds the average payoff of the population, *a fortiori*, the payoff of the mutant. The strategy frequency σ is an ESS if and only if for all $\tau \neq \sigma$, $\mu = 1$ is an asymptotically stable point of (6.3), i.e., there exists $\varepsilon > 0$ such that $\mu_0 > 1 - \varepsilon$ implies $\lim_{t \to \infty} \mu(t) = 1$.

It has been argued that the dynamic based on monomorphic population does not capture one important aspect of the real dynamic process. In a monomorphic population, each individual takes a common mixed strategy. In the real world, however, it is often the case that a population consists of various types of individuals, each of whom takes a pure strategy. In such a case, $\sigma(a)$ is viewed as the fraction of individuals who take the pure strategy a; i.e., σ is a distribution of individuals taking each of the pure strategies. We call such a population a polymorphic population. The dynamic in a polymorphic population is different from that dictated by (6.3). It is called the *replicator dynamics* and is given by

$$\frac{\dot{\sigma}_t(a)}{\sigma_t(a)} = \Pi(\sigma_t; [a]) - \Pi(\sigma_t; \sigma_t), \quad \sigma_0 : \text{ given}, \qquad (6.4)$$

where $\dot{\sigma}_t(a)$ is the rate of change in $\sigma_t(a)$. The second term of the right hand side is the average payoff of the population. An action a increases its frequency if and only if its payoff $\Pi(\sigma_t; [a])$ is higher than the average. Taylor and Jonker (1978) and Zeeman (1981) show that an ESS is an asymptotically stable point of the replicator dynamic, but not vice versa. It is also shown by van Damme (1987) that the game described in Table 6.5 also serves an example of a discrepancy between ESS and asymptotic stability since $\sigma^* = \frac{1}{3}[L] + \frac{1}{3}[C] + \frac{1}{3}[R]$ is an asymptotically stable point of the replicator dynamic.

Evolutionary game theory has overcome a problem of some earlier attempts on learning by considering a large population in which each individual is naturally treated as a price/behavior taker. For example, the study of fictitious play was discouraged when Shapley (1964) presented an example in which the adjustment process

Table 6.5 van Damme's example $(0 < \alpha < 1)$

	L	C	R
L	α, α	$1, -1$	$-1, 1$
C	$-1, 1$	α, α	$1, -1$
C	$1, -1$	$-1, 1$	α, α

does not converge to a Nash equilibrium in a 3-by-3 game, but rather forms a limit cycle. It was discouraged because the very program of fictitious play was to construct an algorithm that converges to Nash equilibrium and hence to justify it. On the other hand, such a cycle is viewed as a natural consequence of the dynamics (Gilboa and Matsui 1991).

There are many other dynamics that have been proposed and analyzed. Sandholm (2010) and Hofbauer and Sigmund (1998) are excellent textbooks for those who are interested in this area of research.

Chapter 7
The More, the Better

This chapter introduces the two key concepts of strategic complementarity and externality, which are related to each other. Take a large train station as an example. People come and go. When observing stairs, two flows of people are usually apparent with one ascending and the other descending. Generally, the flows do not collide with each other but move rather smoothly. People sometimes walk on the left side, while they may walk on the right side at other times. If everyone else walks on the left side, it is beneficial to walk on the left rather than on the right. On the other hand, if everyone else walks on the right side, it is beneficial to walk on the right. In general, the more people walk on the left, the more incentive one has to also walk on the left. We call such a characteristic *strategic complementarity*.

Many customs and conventions exhibit strategic complementarity. When many people shake hands upon meeting someone, it is better to extend your hand for a handshake rather than to make a bow. Conversely, it is somewhat awkward to extend your hand in a society where everyone else bows.

7.1 Scale Economy, Coordination Games, and Strategic Complementarity

In order for a market to function well, we often need scale diseconomy, or diminishing returns to scale. If there is a scale economy, the market may fail to achieve efficiency.

Suppose that each firm has a cost structure that exhibits scale economy:

$$C(x) = F + cx,$$

where $C(x)$ is the total cost of production when the firm produces x units of output, F is the fixed cost that is incurred by the firm whenever it operates in the market,

© Springer Nature Singapore Pte Ltd. 2019
A. Matsui, *Economy and Disability*, Economy and Social Inclusion,
https://doi.org/10.1007/978-981-13-7623-8_7

and $c > 0$ is the marginal cost, or the unit cost of production. In this situation, the average cost of production is given by

$$\frac{C(x)}{x} = \frac{F}{x} + c.$$

As this expression shows, the more output the firm produces, the less becomes the average cost. As a result, the market tends to be occupied by the small number of firms, if not a monopolist.

The same idea can be applied to a disability-related issue. Take, as an example, a ramp for wheelchairs. Once the ramp is constructed, it can be used by many wheelchairs with a negligible cost. Its cost structure is given by

$$C(x) = F + cx,$$

where $C(x)$ is the total cost, F is the cost of constructing the slope in the beginning, and c is the marginal cost, reflecting its maintenance cost per user. Suppose further that the benefit of using the slope is $b < F$. Then the total welfare (total benefit minus total cost) becomes

$$W(x) = bx - (F + c(x)) = (b - c)x - F.$$

Even if the individual benefit b exceeds the marginal cost c, i.e., $b > c$ holds, it may be the case that the total welfare is negative because of the fixed cost F. Also, let the total potential demand for the ramp to be X. To make the analysis interesting, we assume that $W(X) > 0$ holds:

$$X > \frac{F}{b - c}.$$

This means that if everyone uses the ramp, it is beneficial to build it.

The question is whether the ramp should be built in the light of welfare analysis, ignoring the issue of human rights of persons with disabilities for the moment. If the ramp is not built, nobody uses the ramp as a matter of course, and therefore, the welfare is zero. Thus, we should not evaluate the welfare based on the current situation in which nobody in a wheelchair goes out because there is no ramp. We have to estimate how many people would use the ramp if it is ever built. This is often a difficult task because many people simply avoid going out because they feel they are not accommodated.

In order to analyze this situation, let us count the number y of persons with wheelchairs who raise their voice and express their wish to use the ramp. Suppose further that there is a small but positive cost, which may be pecuniary or mental cost, denoted by ε to raise one's voice. Assume $0 < \varepsilon < b$.

The authority builds the ramp if $W(y) > 0$ holds, i.e.,

$$(b - c)y - F > 0,$$

or

$$y > \bar{x} \equiv \frac{F}{b-c}.$$

On the other hand, it does not if $W(y) < 0$ (ignoring the tie situation).

We now analyze this situation. Each person with a wheelchair should decide whether to raise her voice. Take a typical decision maker. If the number y of those who raise their voice is zero, then by raising her voice, this decision maker loses ε, and therefore, she would not raise her voice. Since all the decision makers are symmetric, nobody is willing to raise their voice. This becomes one equilibrium. The same logic holds until y reaches $\bar{x} - 1$.

If, on the other hand, y is $\bar{x} - 1$ except for herself, then she has an incentive to raise her voice. The reason is that by raising her voice, the number reaches $\bar{x}$, the ramp is built, and she obtains the benefit of b, while by not doing so, the number does not reach $\bar{x}$, the ramp is not built, and she does not obtain b.

An interesting phenomenon arises when y exceeds $\bar{x}$. Suppose that this is the case. Then she does not have an incentive to raise her voice since irrespective of her decision, the ramp is built, and therefore, she can save ε by not raising her voice. Once one is accommodated, the activity toward inclusion is toned down.

Summarizing the above analysis, we have found two types of equilibria. The first one is the equilibrium in which nobody raises her voice. The second one is the equilibrium in which exactly $\bar{x}$ people raise their voice.

There are at least two observations that we can make out of this analysis. First, in order to accommodate persons with disability, we need to correctly estimate potential users, or those who obtain benefits from a certain policy. Note that counting the number of people who raise their voice may not be sufficient. Second, the activity is toned down once they are accommodated. This reflects the reality where the total number of members of many disability associations decrease over time once they are accommodated.

This corresponds to the actual story of some disability associations. Consider *Zenkoku-Sekizui-Sonshosha-Rengokai* (Spinal Injuries Japan), one of the larger disability associations in Japan, as an example.[1] This association was established in 1959. Its membership fee is 300 yen, or 3 USD a month. As explained by a *Lohas Medical* newsletter:

> As the result of their success and thanks to their longtime activities, institutions and infrastructures have been formed, and it is now easier to return to society than before, even if one suffers from a cord injury. Consequently, there are more people than ever before who accept the situation as a matter of course. Ironically, the very success of the activities of the association makes one feel that the activities are no longer needed. ... The membership of the association used to be over 5000, but now it is less than 3000.

---

[1] http://lohasmedical.jp/archives/2009/07/post-71.php (access date: April 29, 2018).

Even after a society starts accommodating persons with disabilities, by building ramps and improving infrastructure, there remain many other issues that the society should overcome to accommodate persons with disability. Unless people in the younger generations recognize this situation, it will be difficult for the association to continue its activities.

7.2   Network Externality

Network externality can be explained by using a public transportation system.[2] Suppose that there are n stations along some subway, 1, 2, ..., n. In order to accommodate people using wheelchairs, the stations need to be accessible by, say, elevators. If only one station, say, A is accessible by wheelchairs, it is useless for people using wheelchairs since the purpose of using the underground is to go to another station. Therefore, at least two stations must be accessible by wheelchairs. But again, unless one uses these two stations, say, A and B, there is no use. Thus, only one pattern of going between A and B can be accommodated. What happens if three stations, A, B and C, become accessible? There are now three patterns that can be accommodated. In general, if n stations become accessible by wheelchairs, $n(n-1)/2$ patterns of moves are accommodated. This implies that the accessibility of the underground increases at a faster rate than the number of stations with wheelchair access. This phenomenon, which is already well known, is called network externality.

One may argue that the accessibility of underground stations is not enough for those who wish to move around by themselves. We also need accessibility for buildings, streets, buses, and so forth. Some economists call such a situation *institutional complementarity*. Another notable example can be found in the New York subway. In the 1980s, the New York subway was considered dangerous for passengers. Thefts were common, and the train cars were covered with graffiti. In the late 1980s and early 1990s, there was a dramatic change. To make the subway safe, the number of police officers on board the trains was increased. Measures were also taken so that graffiti could be erased quite easily. As a result, the New York subway became a fairly safe means of transportation.

To see the relationship between the numbers of thefts and police officers, we use an illustrative model. Some people commit thefts if the rate of arrest is sufficiently low. Different (potential) thieves have different thresholds for action. Suppose, for the sake of illustration, that their thresholds are distributed uniformly over the interval of 0 to 1 where 1 means certain arrest. If the rate of arrest is x, which is a number between 0 and 1, then the fraction $1 - x$ of potential thieves are better off committing thefts, while the fraction x had better not. Therefore, the number of thieves is endogenously determined as a function of the arrest rate. However, the number x itself is in turn determined by the number of thieves. Suppose that x is determined by the number $m(> 0)$ of policemen as well as the number n of actual thefts. To simplify the

---

[2]This section is related to Kawagoe and Matsui (2012).

calculation, let $x = m/n$. If n is greater than $1 - x$, then n gradually decreases, while if n is less than $1 - x$, n gradually increases over time:

$$\dot{n} = \alpha(1 - x - n) = \alpha(1 - \frac{m}{n} - n),$$

where $\dot{n}$ is the derivative of n with respect to time; $\dot{n} > 0$ implies that n is increasing, while $\dot{n} < 0$ implies that n is decreasing.

In this dynamics, there are potentially two stable rest points, $n^* = 0$ and $\tilde{n} = [1 + \sqrt{1 - 4m}]/2$, where the second one appears if $1 - 4m > 0$ or $m < 1/4$. There is also an unstable rest point, $\bar{n} = [1 - \sqrt{1 - 4m}]/2$, if, again, $m < 1/4$ holds. If we relax and do nothing, and if n becomes greater than $\bar{n}$ for some reason, then the system moves toward $\tilde{n}$, a bad rest point, or equilibrium.

Its implication is significant. Unlike simple externality, which can be corrected through taxation, the network externality and complementarity pose a big challenge to a market economy: Adam Smith's invisible hand does not work here. Examples are abundant. In Japan, "the collapse of medical services" has been an issue. There was a hospital from which some pediatricians left. This left the last-remaining pediatrician at the hospital in an extremely difficult position, who shortly afterward also resigned.

Similar problems can be seen for care managers and care workers. The compensation for care workers comes from care insurance provided by the government. Therefore, it is not competitively determined but rather set by the government. Because the compensation is low, many care workers quit their jobs within a relatively short time. In many care institutions, the shortage of care workers induces overwork, which further reduces the number of care workers. The lack of a price adjustment mechanism together with the complementarity problem poses a serious problem for the care industry.

7.3 Institutional Complementarity

Our society consists of various institutions.[3] Two major institutions are the market institution and the political institution. The driving forces of institutional development are different between the two. The market institution uses the price adjustment mechanism and profit generation as its driving forces. On the other hand, the political institution uses so-called political power, backed by voters in the case of democratic countries, as its driving force. Since the two institutions use different driving forces, there is no guarantee that the two institutions are harmonized with each other.

Even if the government wishes benevolent so as to make its people as happy as possible, such a goal is not easy to accomplish. Smith (2010) expressed this situation eloquently:

---

[3] See Aoki (2001) for extensive discussion on comparative institutional analysis.

The man of system is nothing like that. He is apt to be sure of his own wisdom, and is often so in love with the supposed beauty of his own ideal plan of government that he can't allow the slightest deviation from any part of it. He goes on to establish it completely and in detail, paying no attention to the great interests or the strong prejudices that may oppose it. He seems to imagine that he can arrange the members of a great society as easily as a hand arranges the pieces on a chess-board! He forgets that the chessmen's only source of motion is what the hand impresses on them, whereas in the great chess-board of human society every single piece has its own private source of motion, quite different from anything that the legislature might choose to impress on it. If those two sources coincide and act in the same direction, the game of human society will go on easily and harmoniously, and is likely to be happy and successful. If they are opposite or different, the game will go on miserably and the society will be in the highest degree of disorder all the time. (Chap. 2, Sect. 2, Part VI)

If there are multiple institutions, we encounter the problem of coordination among them. Using the example of wheelchair access again, if a woman in a wheelchair wants to go to her office by subway, the subway system must accommodate her with elevators, ramps, and general accessibility. At the same time, the street system has to accommodate her by eliminating steps and stone pavements with irregular gaps. In addition, the office building must be designed so that she can move around unimpeded. She cannot go to her office unless all three institutions accommodate her. In this sense, these institutions exhibit strong complementarity in accommodating persons that use wheelchairs.

This is a strategic situation. To accommodate persons using wheelchairs, three institutions have to coordinate. Suppose that they all understand the benefit of a person with a disability being accommodated. Let $b > 0$ be the benefit of accommodation, which is realized only when she can go to the office. Note again that this benefit does not have to be pecuniary one. There is a small cost $c > 0$ of accommodation for each institution. Assume $3c < b$, which implies that it is socially beneficial to accommodate her than otherwise.

In this case, there are multiple equilibria. The first equilibrium is where no institution accommodates persons with disability, while the second equilibrium is where all the institutions accommodate them. The first scenario is an equilibrium because it is against their interests to pay the cost without obtaining the benefit; that is, no party has an incentive to invest for accommodation because nobody else does so. To escape from such an inferior equilibrium, we need a principle or mechanism that is different from unilateral payoff maximization. One solution is to impose restrictions on some party in question, either by law, convention, or social pressure. This approach works for public entities such as railroad companies, but it can be problematic for private business because accommodation might be too costly to be incurred by private business. Reasonable accommodation is the principle that reflects this point.[4]

---

[4]According to the U.S. Equal Employment Opportunity Commission, reasonable accommodation is described in the statement: "The Americans with Disabilities Act (ADA) requires an employer with 15 or more employees to provide reasonable accommodations for individuals with disabilities, unless it would cause undue hardship. A reasonable accommodation is any change in the work environment or in the way a job is performed that enables a person with a disability to enjoy equal employment opportunities.".

Part III
Toward a Theory of Economy and Disability

Chapter 8
Best Response Dynamics

8.1 Introduction

This chapter introduces a dynamic process called the best response dynamics. This is an evolutionary process that does not require full rationality of the game players. In societal phenomena like disability, evolutionary game theory sometimes provides us with a more powerful tool than the standard game theory, which assumes the full rationality of the players as discussed in Sect. 6.2.

In the field of noncooperative game theory, Nash equilibrium (Nash 1951) has played a central role as a solution concept.[1] In bold strokes, one may discern two major interpretations of Nash equilibrium in the context of rational players.

The first, which is close to the "eductive" interpretation of Binmore (1987, 1988), and the "complete information" interpretation of Kaneko (1987), assumes that the game is played exactly once (if it is a repeated game, the repetition occurs once), and the players have sufficient knowledge and ability to analyze the game in a rational manner. Sometimes it is assumed that all players have consistent hierarchies of beliefs, where the game and their priors are common knowledge. Bayesian interpretation such as that proposed by Aumann (1987) advanced this idea to the level that the players have a common prior. From this point of view, however, Nash equilibrium seems far from being satisfactory as it does not satisfy some requirements of "strategic stability." Thus, many studies have been made to refine the concept; among them are Selten (1975), Myerson (1978), Kalai and Samet (1984), and Kohlberg and Mertens (1986). Some studies (see, e.g., Brandenburger and Dekel 1987) loosen the requirement of common knowledge, but still require some a priori knowledge.

The second interpretation is sometimes referred to as the "evolutive" (Binmore) or "naive" interpretation (Kaneko). It does not require that participants in the game know its structure or other facts at the outset. According to this interpretation, a similar situation is repeated many times, and people use trials and errors in choosing

---

[1] This chapter is based on Gilboa and Matsui (1991) and Matsui (1992).

© Springer Nature Singapore Pte Ltd. 2019
A. Matsui, *Economy and Disability*, Economy and Social Inclusion,
https://doi.org/10.1007/978-981-13-7623-8_8

better strategies based on information they gradually acquire. A Nash equilibrium is considered as a stationary point in this repeated situation.

At this point, it is worth noting that the price theory of an earlier age such as Walrasian economics shares the basic view of the world with the naive interpretation. It assumes rational participants in the economy but does not assume any common knowledge among participants. They do not know and do not have to know the entire structure of the economy; rather, they observe aggregated signals such as prices on the basis of which they determine their behavior.

This "naive" price theory has solved many economic problems under some appropriate assumptions on the market structure. For example, in a perfect competition model, the assumption of price-takers results in the participants having (usually unique) dominant strategies as a function of the price signal. The purpose of the present analysis is to apply similar analysis to general n-person normal-form games. In our model, we assume a large population out of which individual players are randomly matched to play a one-shot normal-form game; hence each one of them may consider oneself a "price-taker" and ignore one's effect on others' behavior.

In price theory and game theory alike, there is interest in the stability of an equilibrium, and more generally, in the dynamics of processes that may or may not lead to an equilibrium. However, in our interpretation of a game, this question seems even more relevant and unavoidable than in price theory since Nash equilibrium in mixed strategies typically involves non-unique best responses. To support Nash equilibrium in our interpretation we have to assume that a certain portion of the population chooses each specific strategy, while all the population is indifferent among several of them. In other words, even if all players are perfectly rational and the population is at equilibrium, there is no compelling reason to believe it would stay there. There are equally or more probable scenarios according to which every individual plays optimally and yet the behavior pattern moves away from the equilibrium point.

In defining a solution concept on the basis of the naive interpretation, we require it to satisfy the following four qualifications. First, as in a perfectly competitive market, it is assumed that each player is sufficiently small and anonymous, and then may maximize her expected utility without getting involved in complicated strategic considerations such as retaliation. Second, unlike a deviation made by a single player, a change in behavior pattern is made in a continuous way. This expresses the intuitive idea that within a short time interval only a correspondingly small proportion of the individuals realize the current behavior pattern and change their strategies. Third, individuals are myopic and choose best response strategies to the current behavior pattern. The important consequence of this assumption is that the behavior pattern may form a cycle. Finally, there is a certain limitation in recognizing the current situation. No matter how much information one gathers, it is hard to tell the exact behavior pattern of the society at a given moment.

Similar to the case of complete information, the concept of Nash equilibrium is not satisfactory as a solution concept when we take the above features into consideration. For example, in the game of coordination, which is shown in Table 8.1, there are three Nash equilibria, namely, $([L], [L]), ([R], [R])$, and $(\frac{1}{2}[L] + \frac{1}{2}[R], \frac{1}{2}[L] + \frac{1}{2}[R])$. In the "real world," if the behavior pattern fluctuates toward, say, $([L], [L])$ from the third

Table 8.1 Pure coordination game

	L	R
L	1, 1	0, 0
R	0, 0	1, 1

equilibrium, and if that tendency is observed, then people are likely to follow that behavior. Therefore, the mixed strategy equilibrium of this example is unlikely to sustain itself as a stationary point of some dynamic process.

8.2 Social Stability

This section defines a solution concept, called socially stable set, to capture the dynamic properties, and extend them beyond the mere classification of Nash equilibria to stable and unstable ones.

We first consider the following notion of accessibility, the precise definition of which will be given in the following section: given $\varepsilon > 0$, a strategy profile g is ε-accessible from f if there is a continuous path starting with f and ending in g, such that the direction at each point of the path is a best response to some strategy in the ε-neighborhood of that point; a strategy profile g is accessible from f if there exists a g' sufficiently close to g and ε sufficiently close to zero such that g' is ε-accessible from f. A socially stable set is a set of strategy profiles such that no strategy profile outside the set is accessible from any strategy profile inside the set, and all the strategy profiles in the set are accessible from each other. In particular, if the socially stable set is a singleton, we call its element a socially stable strategy. We will prove that socially stable sets always exist and that each one of them is closed and connected. This new, set-valued solution concept is quite different from various refinements of Nash equilibrium, such as trembling hand perfect (Selten 1975), persistent (Kalai and Samet 1984), proper (Myerson 1978), evolutionarily stable strategy (Maynard Smith and Price 1973), and Kohlberg and Mertens' stable set (Kohlberg and Mertens 1986). Socially stable set as a solution concept differs from the notion of fictitious play although the best response dynamics that is used to define accessibility is essentially equivalent to fictitious play. We may say that socially stable set is a solution concept based on either best response dynamics or fictitious play (see Footnote 2 on page 74).

One important feature of socially stable sets is the independence of sequential elimination of strictly dominated strategies. The situation we have in mind is that all the individuals are so "small" that they do not have to consider the effect of their choices on the distribution of the population, and that all the individuals make no mistakes except that they cannot recognize the present situation precisely (even in that case, their choices are made in a rational manner on the basis of their observation.) In this situation no one should care about strictly dominated strategies, which cannot be chosen at a stationary state. On the other hand, weakly dominated strategy may be present in the support of strategy profiles in a socially stable set since an individual does not care or does not even know the payoff difference that appears only when other types of individuals take strategies that are not used. Note that Selten's concept of trembling hand perfectness and Kalai and Samet's persistent equilibrium are affected by strictly dominated strategies.

Another important property is independence of redundant strategies. Note that Myerson's proper equilibrium does not satisfy this property.

Our model and solution concept are general enough to deal with various random-matching processes. Consider for simplicity a game in which two people are matched. Then the following two cases are distinguished. In the first, the two people matched are from different groups of individuals, say, male and female. In the second, they belong to the same type. In n-person games, in which there are exactly n participants, this distinction is irrelevant since each person is assumed to have her own identity. On the other hand, in n-type games, which typically involve many participants of each type, information is gathered about types, while the decision makers are individuals. Hence, should two individuals of the same type be matched, each may choose a strategy independently of the other, but the aggregate strategy profile must be symmetric. We allow the model to cope with both situations. Our results are stated and proved in a general framework in which a "game" involves the encounter of several (possibly one) individuals of each type. The socially stable sets will, of course, depend on the assumptions regarding the identity of types of different players.

8.2.1 Definitions and Notations

In a society, as in a "game," there are several types of individuals. Some people are matched randomly to take some actions. In each matching situation, the number of participants from each type is fixed and may exceed one. Therefore, depending on the setting, two individuals of the same type may be matched.

Formally, a game G is described as a quadruple:

$$G = \langle I, M, (S_i)_{i \in I}, (\pi_i)_{i \in I} \rangle.$$

where $I = \{1, 2, \ldots, n\}$ is the set of types of individuals, S_i ($i \in I$) is the finite set of strategies for each individual of type i, $M = (m_1, m_2, \ldots, m_n)$ specifies the number of individuals of each type who are matched in each matching situation, and $\pi_i : \times_{j \in I} S_j^{m_j(i)} \times S_i \to \mathbb{R}$ where $m_i(i) = m_i - 1$ and $m_j(i) = m_j$ if $j \neq i$ is a payoff function for each individual of type i, where a typical value $\pi_i(s_1^1, \ldots, s_1^{m_1}, \ldots, s_i^1, \ldots, s_i^{m_i - 1}, \ldots, s_n^{m_n}; s_i)$ is the payoff for individual of type i when he takes s_i, while others take $(s_1^1, \ldots, s_n^{m_n})$. This somewhat awkward definition of the domain will simplify notation in the sequel.

We assume that π_i is invariant with respect to permutation of strategies among the same type, i.e., among $s_j^1, \ldots, s_j^{m_j(i)}$. We bear in mind the interpretation according to which each $i \in I$ consists of a sufficiently large number of individuals who are anonymous and are matched randomly in each instance; without this interpretation, the definitions in the following sections will have little validity. Let $\mathcal{F}_i \equiv \Delta(S_i)$ be the set of probability distributions over S_i, i.e.,

$$\mathcal{F}_i \equiv \Delta(S_i) = \left\{ f_i : S_i \to \mathbb{R} \,\middle|\, \sum_{s_i \in S_i} f_i(s_i) = 1, f_i(s_i) \geq 0 \text{ for all } s_i \in S_i \right\}.$$

We may call $\mathcal{F} \equiv \times_{i \in I} \Delta(S_i)$ the class of strategy profiles and $f \equiv (f_1, \ldots, f_n) \in \mathcal{F}$ a strategy profile. In considering the dynamic adjustment process, the current strategy profile will often be referred to as a behavior pattern. $\mathcal{F}$ is considered as $(\sum_{i \in I} |S_i| - n)$-dimensional space on which Euclidean norm, $\| \cdot \|$, and linear operations are defined.

Given a strategy profile $f \in \mathcal{F}$, the expected payoff for an individual of type i ($i \in I$) if he takes a strategy $r_i \in S_i$ is:

$$\Pi_i(f; r_i) = \sum_{s \in \times_{j \in I} S_j^{m_j(i)}} \prod_{j \in I} \prod_{k=1}^{m_j(i)} f_j(s_j^k) \pi_i(s; r_i).$$

Let $Br_i(f)$ be the set of pure strategies for individuals of type $i \in I$ that are the best responses to f, i.e.,

$$Br_i(f) = \arg \max_{r_i \in S_i} \Pi_i(f; r_i).$$

We let $BR_i(f) \equiv \Delta(Br_i(f))$, $Br(f) = \times_{i \in I} Br_i(f)$, and $BR(f) = \times_{i \in I} BR_i(f)$. Given $F \subset \mathcal{F}$, we denote $Br_i(F) \equiv \cup_{f \in F} Br_i(f)$.

Let a function $[\cdot] : S_i \to \Delta(S_i)$ ($i \in I$) satisfy $[s_i](s_i) = 1$ for all $s_i \in S_i$. The ε-neighborhood of a strategy profile f, denoted by $U_\varepsilon(f)$, is the set of strategy profiles g the distance of which from f in the Euclidean norm is less than ε.

8.2.2 Socially Stable Sets

This subsection defines and discusses the concepts of socially stable set and socially stable strategy. First, the definition of Nash equilibrium is given.

Definition 8.1 A strategy profile $f^* \in \mathcal{F}$ is a Nash equilibrium if f^* is a best response to f^* itself, i.e., $f^* \in BR(f^*)$.

To capture the idea of social stability, we consider the following three points: (1) there are no strategic considerations such as retaliation; (2) unlike a deviation made by a single player, a change in behavior pattern is likely to be continuous; and

(3) each player's ability to recognize the current situation is limited. To express these points, we introduce the notion of ε-accessibility.

Definition 8.2 Given $\varepsilon > 0$ and strategy profiles f and g, g is ε-accessible from f if there exist a continuous function $p : [0, 1] \rightarrow \mathcal{F}$ differentiable from the right, a function $b : [0, 1] \rightarrow \mathcal{F}$ continuous from the right, and $\alpha \in [0, \infty)$ such that

$$p(0) = f, \quad p(1) = g,$$

and for each $t \in [0, 1)$,

$$\frac{d^+}{dt} p(t) = \alpha[b(t) - p(t)], \text{ and}$$

$$b(t) \in \times_{i \in I} BR_i[U_\varepsilon(p(t))].$$

The definition says that in the case of $\alpha > 0$, a behavior pattern moves in the direction of a convex combination of best responses to some strategy profiles that are in the ε-neighborhood of the behavior pattern, and it stays at the same place only if the behavior pattern is a best response to another one that is in the ε-neighborhood of itself. By including the case of $\alpha = 0$, we assure that a strategy profile is always ε-accessible from itself.[2]

The interpretation of this definition is that only small and equal portions of individuals in each type realize the current behavior pattern and change their behavior pattern to another which is a best response to it. In doing so, there is a limitation on the ability of recognizing the current behavior pattern, so that its change may not be directed toward a best response to it; rather, it is only assumed that the direction is a best response to a possibly different behavior pattern that is in the ε-neighborhood of the current one. We may call the function p an ε-accessible path from f to g. Using this, accessibility from one strategy profile to another is defined.

Definition 8.3 For two strategy profiles f and g, g is accessible from f if there exist sequences $(\varepsilon_n)_{n=1}^{\infty}$ in $(0, \infty)$ and $(g^n)_{n=1}^{\infty}$ in $\mathcal{F}$ convergent to 0 and g respectively such that g^n is ε_n-accessible from f for all n.

Now, we are in a position to present the definition of social stability.[3]

---

[2]The unperturbed version of this dynamic process has been developed as *fictitious play*. The mathematical definitions of the unperturbed best response dynamics and the fictitious play are identical even though their setups are very different: the former considers a continuum of players who take myopic best responses against the present strategy distribution, while the latter considers finitely many players who take best responses against the past average of the opponents' behavior. Earlier contributions concerning fictitious play include Brown (1951), Miyasawa (1961), and Shapley (1962). For more recent development, see, e.g., Monderer and Shapley (1996) and Hofbauer and Sandholm (2002). Fudenberg and Levine (1998) presents a concise introduction to both fictitious play and best response dynamics.

[3]This concept was called "cyclically stable set" in Gilboa and Matsui (1991).

Definition 8.4 A nonempty subset F^* of $\mathcal{F}$ is a socially stable set (SS set) if no $g \notin F^*$ is accessible from any $f \in F^*$, and every $f^* \in F^*$ is accessible from all f in F^*.

A strategy profile $f^* \in \mathcal{F}$ is called a socially stable strategy (SS strategy) if $\{f^*\}$ is a socially stable set as a singleton.

A socially stable set is stable in the sense that once the actual behavior pattern falls into it, another strategy profile may be realized if and only if it is within the socially stable set. The interpretation of this concept is as follows. For a long time, individuals have sought better strategies. After they search all the alternatives and acquire almost complete knowledge about the behavior pattern of other individuals, the actual behavior pattern may move within a socially stable set but never leave it. The paths may be complicated, especially when there are tie situations, in which case the behavior pattern may fluctuate arbitrarily along a continuum of strategy profiles. Before we present the properties of socially stable sets, we present some important properties of the notion of accessibility, which are summarized in the following two lemmata.

The first lemma is closedness with respect to accessibility.

Lemma 8.1 *Suppose that $(g^n)_{n=1}^{\infty}$ is a sequence of strategy profiles all of which are accessible from $f \in \mathcal{F}$. If (g^n) converges to $g \in \mathcal{F}$, then g is accessible from f.*

Proof Let there be given $(g^n)_{n=1}^{\infty}$, f, and g as above. For each g^n, there exists a sequence (g^{nk}) such that g^{nk} is in the $\frac{1}{k}$-neighborhood of g^n and is $\frac{1}{k}$-accessible from f. Take the diagonal sequence $(\mu^k) = (g^{kk})$. Then (μ^k) converges to g, and μ^k is $\frac{1}{k}$-accessible from f. Thus, g is accessible from f. $\qquad\square$

The next lemma is transitivity.

Lemma 8.2 *If h is accessible from g, which in turn is accessible from f, then h is accessible from f.*

Proof Suppose that h is accessible from g and that g is accessible from f. Then there exists a sequence (g^n) converging to g such that g^n is $\frac{1}{n}$-accessible from f. Given $\delta > 0$, there exists $\bar{n}$ such that $g^n \in U_\delta(g)$ holds for all $n > \bar{n}$. Since h is accessible from g, there exists a δ-accessible path from g to $g' \in U_\delta(h)$, denoted by p. We construct a 2δ-accessible path from g^n to $g'' \in U_{2\delta}(h)$, denoted by q, by using p. Since p is a δ-accessible path from g to g', p is a solution to the problem:

$$\frac{d^+}{dt}p = \alpha_0(b^0 - p), \quad p(0) = g,$$

for some $\alpha_0 \geq 0$ and a function b^0 continuous from the right on $[0, 1]$. Since b^0 is continuous from the right, it has no more than a countable number of discontinuity points. Consider the problem: find a continuous q such that

$$\frac{d^+}{dt}q = \alpha_0(b^0 - q), \quad q(0) = g^n.$$

By a well known theorem (see, e.g., Coddington and Levinson 1955, pp. 75–78), such a q exists and is unique. Moreover, since b^0 is continuous from the right, $\frac{d^+}{dt}q$ equals $\alpha_0(b^0 - q)$ even at the discontinuity points of b^0.

Now, since $\|p(0) - q(0)\| < \delta$ holds, and p is a δ-accessible path, it is sufficient to show that $\|p(t) - q(t)\|$ is nonincreasing in t. If $\alpha_0 = 0$, the claim trivially holds, so suppose $a_0 > 0$. First, we have

$$\frac{d^+}{dt}(p - q) = \alpha_0(b^0 - p) - \alpha_0(b^0 - q) = -\alpha_0(p - q).$$

Then we have

$$\|p(t + \tau) - q(t + \tau)\| \le \left\| \{p(t) - q(t)\} + \frac{d^+}{dt}(p(t) - q(t))\tau \right\| + o(\tau)$$
$$= \|(1 - \alpha_0\tau)\{p(t) - q(t)\}\| + o(\tau),$$

which is smaller than $\|p(t) - q(t)\|$ for a sufficiently small $\tau > 0$. Thus, there exists $g'' \in U_{2\delta}(h)$ which is η-accessible from f where $\eta = \max(2\delta, 1/n)$. This is true for all $n > \bar{n}$, and δ is arbitrary. Therefore, h is accessible from f.   □

8.2.3   Properties of Socially Stable Sets

In this section, we prove that socially stable sets exist. Also, we will see the relationship between Nash equilibrium on the one hand and socially stable set and socially stable strategy on the other.

Existence

Before we state and prove the existence theorem for a socially stable set, we denote by $R(f)$ the set of strategy profiles that are accessible from f, i.e., given $f \in \mathcal{F}$,

$$R(f) = \{g \in \mathcal{F} \mid g \text{ is accessible from } f\}.$$

In the proof, we make use of Zorn's lemma and the lemmata presented in the previous section.

Theorem 8.3 *Every game has at least one socially stable set.*

Proof First, observe that $R(f)$ is nonempty for any $f \in \mathcal{F}$, that, by Lemma 8.1, $R(f)$ is closed for any f, and that, by Lemma 8.2, $f' \in R(f)$ implies $R(f') \subset R(f)$.

Next, we consider the family of sets $\{R(f)\}_{f \in \mathcal{F}}$ and define the inclusion $\subset$ as a partial order on them. Take any family $\{f^\alpha\}_{\alpha \in A}$ of strategy profiles such that for any α and β in A, either $R(f^\alpha) \subset R(f^\beta)$ or $R(f^\beta) \subset R(f^\alpha)$ holds. Consider $\cap_{\alpha \in A} R(f^\alpha)$, which is nonempty since the $R(f^\alpha)$'s are compact. Choose any $f \in \cap_{\alpha \in A} R(f^\alpha)$ and recall that $R(f) \subset R(f^\alpha)$ holds for all $\alpha \in A$. Hence, $R(f)$ is a lower bound of the

$R(f^\alpha)$'s. Therefore, by Zorn's lemma, there exists a minimal element $R^* = R(f^*)$ among the $R(\cdot)$'s. It is not empty because all the sets $R(f)$'s are nonempty.

We now claim that R^* is a socially stable set. Indeed, for any $f \in R^*$, Lemma 8.2 implies $R(f) \subset R^*$. On the other hand, $R^* \subset R(f)$ holds for any $f \in R^*$ since R^* is a minimal element. Thus, $R(f) = R^*$ holds, which implies that every point in R^* is accessible from any point in R^*, and no point outside R^* is accessible from any point in R^*. □

By a similar argument, we can prove that for any strategy profile f, there exists a socially stable set any element of which is accessible from f. That is to say, the "domain of attraction" of all the socially stable sets is the whole space of mixed strategies (where a point f is said to be attracted to a socially stable set F^* if there exists $g \in F^*$ that is accessible from f; obviously, f may be attracted to several socially stable sets). It is also worth noting that every socially stable set and its domain of attraction are closed and connected, that it is invariant with respect to sequential elimination of strictly dominated strategies and redundant strategies. We also note here that socially stable sets are neither upper nor lower hemi continuous with respect to the game payoffs.

Nash Equilibrium and Social Stability

We first have the following proposition.

Proposition 8.4 *Any socially stable strategy is a Nash equilibrium.*

Proof Suppose that a strategy profile f is not a Nash equilibrium. Then there exist $\delta > 0$ and $s_i \in S_i$ for some $i \in I$ such that any strategy profile in $U_\delta(f)$ takes s_i with probability of at least δ and $s_i \notin Br_i(U_\delta(f))$ holds. Then for any $\varepsilon > 0$, there exists an ε-accessible path p that reaches the boundary of $U_\delta(f)$ since the speed of decrease in $p_i(t)(s_i)$ is positive and bounded away from zero. Thus, there is a strategy profile in the boundary of $U_\delta(f)$ that is accessible from f since the boundary is sequentially compact. Hence, f cannot be a socially stable strategy. □

Next, we define a strict Nash equilibrium as a strategy profile f^* such that $BR(f^*) = \{f^*\}$, i.e., f^* is a profile of strategies that are strictly better responses to f^* than any other strategy. Then any strict Nash equilibrium is a socially stable strategy since for a sufficiently small $\varepsilon > 0$, the set of the best response directions consists only of itself. Note that the converse is not true in general. In the game of matching pennies, for example, the mixed strategy Nash equilibrium is a socially stable strategy; on the other hand, it is not a strict Nash equilibrium (recall that any mixed strategy profile cannot be a strict Nash equilibrium).

The concept of a socially stable set is not directly related to that of Nash equilibrium. Although a socially stable strategy is always a Nash equilibrium, each Nash equilibrium may be in some socially stable set or outside any of the socially stable set. We proceed to show an example of a game that has no Nash equilibrium inside any socially stable set.[4] Consider a one-type game with two individuals matching in

---

[4]Shapley (1962) gives an example where fictitious play does not converge to a Nash equilibrium.

Table 8.2 A game where no nash equilibrium is in any socially stable set

	L	M	R
L	2, 2	1.2, 1.2	−1, 3
M	1.2, 1.2	1, 1	0.2, 0.2
R	3, −1	0.2, 0.2	0, 0

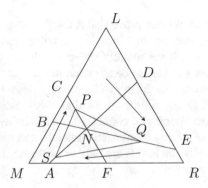

Fig. 8.1 Phase diagram

Table 8.2. This is a modified version of Shapley's example for fictitious play. This game has a unique Nash equilibrium, $(\frac{1}{4}[L] + \frac{1}{2}[M] + \frac{1}{4}[R], \frac{1}{4}[L] + \frac{1}{2}[M] + \frac{1}{4}[R])$ if we regard it as a two-person game. In the following, we let (p, q, r) stand for $(p[L] + q[M] + r[R])$. We will find a socially stable set and then show that it is accessible from the unique Nash equilibrium, which does not belong to it. This will also prove that the Nash equilibrium does not belong to any other socially stable set. Figure 8.1 shows the simplex of strategy profiles. In this figure, the vertex L of the triangle stands for the strategy profile $[L]$ and so on. The line segment AD indicates that if a strategy profile is on this line, then the pure strategies L and M yield the same expected payoff to the individuals. Similarly, on BE, individuals are indifferent between M and R, and on CF they are indifferent between L and R. Therefore, the area $AMCN$ is the one in which players prefer to take L; $CLEN$ is for R; and $ERAN$ is for M. Finally, N is the Nash equilibrium.

In this game, a behavior pattern that differs from N swirls around it indefinitely without reaching any pure strategy profile. In fact, one may find that if the behavior pattern is inside PQS of Fig. 8.1, then it follows an expanding cycle converting to PQS, and if it is outside PQS, it follows one shrinking to PQS, where $P = (.4, .5, .1)$, $Q = (.16, .2, .64)$, and $S = (.04, .8, .16)$. If the behavior pattern is on PQS, then an ε-accessible path remains in some band around PQS. Therefore, PQS is a socially stable set. Since PQS is accessible from N no matter how small ε may be, there is no socially stable set that contains the Nash equilibrium in this game.

Unlike fictitious play, which is developed to find a Nash equilibrium, we do not view this phenomenon as a flaw of the concept of a socially stable set; rather, it seems to us to criticize the Nash equilibrium concept. To the extent that one finds the dynamic process presented above as reasonable, one is led to believe that Nash equilibrium may not be the appropriate tool for analysis of the evolution of economic behavior in large populations.

8.3   Static and Dynamic Concepts of Social Stability

8.3.1   Static versus Dynamic Concepts

An evolutionarily stable strategy (ESS) is a strategy distribution that is robust against an invasion of any small number of mutants. This concept was introduced by Maynard Smith and Price (1973) in order to cope with the question of evolutionary stability. It is simple, yet powerful, in some applications. For example, using modifications, Fundenberg and Maskin (1990) and Binmore and Samuelson (1992) predict efficient outcomes in infinitely repeated games. Although their results are interesting, it is difficult to justify the use of variations of ESS without considering the underlying dynamics. Some dynamic stability concept is needed. Indeed, many attempts have been made to relate ESS to dynamic behavior or to find another static expression for dynamic stability (see the survey by van Damme 1991).

The problem that stimulates these studies is that ESS, although capturing the intuitive idea of evolutionary stability, corresponds only to a restrictive form of dynamics in which all the population take the same mixed strategy and only a single type of mutants (who may also play a mixed strategy) is allowed to enter. In fact, Taylor and Jonker (1978) shows that an ESS is a sufficient but not a necessary condition for "asymptotic stability" in a polymorphic population, which consists of several types of genes, each of which takes a pure strategy.

The previous section of this chapter proposed a solution concept called the socially stable set, which is directly derived from an adjustment process. The basic concept is accessibility, which is roughly defined as follows. A strategy distribution g is accessible from another strategy distribution f if there is a path from f to g where the direction of the path at each point is a best response to that point. A socially stable set is a set of strategy distributions that is closed under accessibility and in which any two members are accessible from each other. Unlike equilibrium theory, which is valid only when players play according to an equilibrium, the notion of a socially stable set does not presume that the society stays in a socially stable set. Modifying the concept of accessibility and weakening the concept of ESS, the present analysis shows the equivalence between a socially stable set of points of a dynamic process and an ESS-like static solution concept.

The reason that we have to weaken ESS to get the equivalence result in a polymorphic population is that ESS does not consider the role played by a third party. A strategy distribution is not an ESS whenever there is a mutant that fares better than

Table 8.3 A rock-scissors-paper game

	R	S	P
R	α, α	$1, -1$	$-1, 1$
S	$-1, 1$	α, α	$1, -1$
P	$1, -1$	$-1, 1$	α, α
			$(-1 < \alpha < 1)$

an individual taking the original strategy. But what happens if there exists another mutant that dominates the initial mutant?

In order to elucidate the point, consider the symmetric game expressed in Table 8.3 (van Damme 1991). The unique symmetric Nash equilibrium of this game is the mixed strategy $p = (\frac{1}{3}, \frac{1}{3}, \frac{1}{3})$, if α is between zero and one. It is not an ESS, however, since strategy $q = (1, 0, 0)$ fares better than p in a population consisting of a large fraction of p and a small fraction of q; namely, $\Pi(p; q) = \Pi(p; p)$, and $\Pi(q; q) > \Pi(q; p)$, where $\Pi(p; q)$ is the payoff of an individual taking q when the distribution of strategies is p.

Here, the comparison is made between two strategy types, p and q. If we view p as a population consisting of the three pure strategies with equal frequencies, then the story is very different. To see this point, let $(1 - \lambda)p + \lambda q$ be a population consisting of the first, second, and third pure strategies with the frequencies of $\frac{1}{3}(1 - \lambda) + \lambda$, $\frac{1}{3}(1 - \lambda)$, and $\frac{1}{3}(1 - \lambda)$, respectively. In this population, the first action R is no longer the fittest. Indeed, the frequency of the third action P starts growing most rapidly. After the frequency of P has grown enough, then the frequency of S will grow. In this way, the population dynamics show a cyclical pattern, and it is no longer obvious that the behavior pattern moves away from the initial distribution.

The dynamics considered in the literature on evolutionary stability is sometimes called replicator dynamics since it reflects the idea that each individual reproduces offspring based on its fitness. In the process, a genotype (or phenotype depending on the context) that fares better than the average increases its frequency, and vice versa.

On the other hand, the best response dynamics is a different type of dynamics. In this process, the frequency of an action increases only if it is a best response to the present strategy distribution. The intuition behind this dynamics is that in a small amount of time, only a small fraction of people die and are replaced by newly born children. These children inherit knowledge of the behavior pattern of the society from their parents and make optimal decisions under a static expectation (namely, without caring about the future change in the behavior pattern). Moreover, once one chooses an action, he sticks to it until his death, because of, for example, a large adjustment cost. We define a socially stable strategy (with respect to the best response dynamics), or SSS(BR), as a strategy from which no best response dynamic path moves away.

We will give a static expression of the "stable" strategy, which requires the robustness of an invasion upsetting the original distribution. While the formal argument will be made in the subsequent sections, the informal description of our stability notion is given as follows: a strategy distribution is a socially stable strategy (against equilibrium entrants), or an SSS(EE), if whenever there is a cluster of mutants, say, cluster 1, whose payoffs are higher than those in the original population, then there exists another cluster, say, cluster 2, which obtains a higher payoff than cluster 1 in a population consisting of a large fraction of the original population and a small fraction of cluster 1. This concept was first introduced by Swinkels (1992) who calls such a strategy profile robust against equilibrium entrants.

The term "equilibrium entrants" stems from the fact that any entrant that upsets the population has to constitute an equilibrium in the game restricted to the set of best response strategies to the original population. It turns out that the two definitions of social stability are equivalent.

Socially stable strategies may not exist, but this is not a drawback to the concept; rather, it is the natural consequence of the dynamic process we investigate. However, one may ask the following: what happens to the actual behavior pattern? If the process is not in a steady state, where does it go? In order to answer these questions, the previous section proposed a set-valued solution concept called a socially stable set. The idea is to include two distinct strategy distributions in the same socially stable set of points if the two points are accessible from each other, that is to say, there are best response paths to and from each other. We give a static expression of the modified version of this set-valued concept as well, which enables us to analyze many situations in a much simpler way. We also show the existence of a socially stable set.

The set-valued static condition derived in the present analysis is a modification of an equilibrium evolutionarily stable set (EES set) proposed by Swinkels (1992). The main difference between the two concepts is that socially stable set has no condition directly related to Nash equilibrium, while an EES set is always a subset of the set of Nash equilibria by definition. We will subsequently compare the two concepts.

8.3.2 Definitions and Notations

In a society, or equivalently a "game," there are several types of individuals. Some people are matched randomly to take some actions. In each matching situation, the number of participants from each type is fixed and may exceed one. Therefore, depending on the setting, two individuals of the same type may be matched.

We use the same notations as those in Sect. 8.2, i.e., a game G is described as a quadruple:

$$G = \langle I, M, (S_i)_{i \in I}, (\pi_i)_{i \in I} \rangle.$$

The description of games is general enough to accommodate various situations. For example, a game with a single population discussed in the literature on evolutionary biology is expressed as

$$G = \langle \{1\}, 2, S_1, \pi_1 \rangle, \tag{8.1}$$

while if one is interested in a random matching situation between male and female, one may have

$$G = \langle \{1, 2\}, (1, 1), (S_1, S_2), (\pi_1, \pi_2) \rangle. \tag{8.2}$$

We bear in mind the interpretation according to which each type consists of a sufficiently large number of individuals who are anonymous and are matched randomly at every instant. Without this interpretation, the following definitions will have little validity.

8.3.3 Evolutionarily Stable Strategies and Replicator Dynamics

As before, the definition of Nash equilibrium is given.

Definition 8.5 A strategy profile $f^* \in \mathcal{F}$ is a Nash equilibrium if f^* is a best response to f^* itself, i.e., $f^* \in BR(f^*)$.

Evolutionarily stable strategy is a concept proposed by Maynard Smith and Price (1973). It is defined for the single population games given by (8.1). Its definition is presented below.

Definition 8.6 Suppose that the game is expressed as (8.1). Then a strategy profile $f^* \in \mathcal{F}$ is an evolutionarily stable strategy (ESS) if for all $g \in \mathcal{F}$,

$$\Pi(f^*; f^*) \geq \Pi(f^*; g), \tag{8.3}$$
$$\Pi(g; f^*) > \Pi(g; g) \quad \text{if } \Pi(f^*; f^*) = \Pi(f^*; g). \tag{8.4}$$

For a game of the form (8.1), (8.3), the first condition of ESS, is equivalent to Nash equilibrium. What makes this concept unique is (8.4). An interpretation is as follows. The original population is f^*. Then a small group of mutants g comes in. The question is whether this group of mutants will flourish as time goes by or not. In order for the original population to prevent the mutants from increasing, the original population fares at least as well as the mutants against the original population [see (8.3)]. Also, in case of a tie against the original population, it must be the case that the original population dominates the mutants against the mutant population [see (8.4)].

ESS is a static concept that captures the stability of some dynamic process. A question arises as to whether or not ESS actually corresponds to stable points in some dynamics. In this subsection, we consider replicator dynamics and a stability concept to see under what condition ESS coincides with the stability of the replicator dynamics.

Consider a game of the form (8.1) with $S = \{s_1, \ldots, s_K\}$. Time is continuous, and let $p : [0, \infty) \to \mathcal{F}$ be a path. In this expression, $p_k(t)$ is the fraction of the agents taking s_k ($k = 1, \ldots, K$) at time t. Then we have the following definition.

Definition 8.7 Given a strategy profile $f_0 \in \mathcal{F}$, a replicator dynamics starting with f_0 is given by a path $p : [0, \infty) \to \mathcal{F}$ that satisfies $p(0) = f_0$ and

$$\dot{p}_k(t) = p_k(t) \left[\Pi(p(t); p_k(t)) - \Pi(p(t); p(t)) \right].$$

In the definition $\dot{p}$ is the derivative of p with respect to time t.

Next, we define the following stability concept. In the definition, we use the concept of neighborhood of a strategy distribution f^*, which is defined to be an open set that contains f^*.

Definition 8.8 A strategy distribution f^* is asymptotically stable if for any neighborhood U of f^*, there exists another neighborhood V of f^* such that for all $f_0 \in V$, the replicator dynamic path p that starts with f_0 never leaves U, i.e., $p(t) \in V$ holds for all $t \geq 0$.

It has been shown that ESS coincides with asymptotically stable points of the replicator dynamics if there are only two strategies, i.e., $|S| = 2$. However, as we have argued in Sect. 8.3.1, ESS is stronger than asymptotic stability in a game with $|S| > 2$.

8.3.4 A Point-Valued Solution Concept and Its Static Equivalence

This subsection first considers the best response dynamics. Given $f_0 \in \mathcal{F}$, a best response dynamic path starting from f_0 is a continuous function p from some time interval $[0, T]$ to the set of strategy distributions that moves toward the best response strategy b at each point in time. The formal definition is given below.

Definition 8.9 Given a strategy profiles $f_0 \in \mathcal{F}$ and $T \in (0, \infty)$, a continuous function $p : [0, T] \to \mathcal{F}$ is a best response dynamic path starting from f_0 if p is differentiable from the right, $p(0) = f_0$, and there exists a step function $b : [0, 1) \to \mathcal{F}$ continuous from the right such that

$$\frac{d^+}{dt} p(t) = b(t) - p(t),$$

and

$$b(t) \in BR(p(t))$$

hold for all $t \in [0, T)$.

This definition implies that the behavior pattern may go only in the direction of the present best response. In this sense, this definition is close to but different from the path defined in Sect. 8.2, which allows some perturbation.

Table 8.4 A game with a unique strongly Pareto optimal outcome

	L	R
L	1, 1	0, 0
R	0, 0	1, 1

Several remarks are in order. First, we always consider the domain of a path to be finite, i.e., a process finishes within a finite amount of time. This restriction is important since we concatenate two paths to construct a new one. Second, we choose h to be a step function in order to avoid bizarre situations like the one in which the frequency of change in the direction of the behavior pattern goes to infinity as t tends to zero. This restriction plays an important role in proving the subsequent results. Third, if f is a Nash equilibrium, then p with $p(t) = f$ for all $t \in [0, T]$ is a dynamic path since f is always a best response to itself. Note also that there may be multiple dynamic paths starting from the same point because of the possible multiplicity of best responses.

We repeat a point-valued solution concept defined in Sect. 8.2.

Definition 8.10 A strategy distribution $f^* \in \mathcal{F}$ is a socially stable strategy (with respect to the best response dynamics), or SSS(BR), if there exists no best response dynamic path $p : [0, T] \to \mathcal{F}$ with $p(0) = f^*$ and $p(t) \neq f^*$ for some $t \in (0, T]$.

An SSS(BR) is defined to be a distribution from which no best response path will depart. It will be shown that an SSS(BR) is a Nash equilibrium. The converse, however, is not always true. Consider a single type society, characterized by (8.1), with two actions and the payoff matrix given in Table 8.4.

Here, although $\frac{1}{2}[L] + \frac{1}{2}[R]$ is a Nash equilibrium, it is not an SSS(BR) since, for example, $p : [0, T] \to \mathcal{F}$ with $p(t) = (1 - e^{-t})[L] + e^{-t}(\frac{1}{2}[L] + \frac{1}{2}[R])$ is a dynamic path departing from $\frac{1}{2}[L] + \frac{1}{2}[R]$. Note that any strict Nash equilibrium is an SSS(BR). Indeed, if $p(0) = f$ is a strict Nash equilibrium, i.e., f is the unique best response to itself, then $d^+p/dt = b(t) - p(t) = 0$ always holds since $b(t) \in BR(p(t)) = \{f\}$ holds if $p(t) = f$.

Next, we provide a static definition that captures the dynamic stability of a point. It says that a distribution f is "stable" if for any distribution g that is a best response to f, there is another distribution h that fares better than g in a population of $(1 - \varepsilon)f + \varepsilon g$ with a sufficiently small $\varepsilon > 0$.

The formal definition is given in the following, which was first proposed by Swinkels (1992).

Definition 8.11 (Swinkels 1992) A strategy distribution $f^* \in \mathcal{F}$ is a socially stable strategy (against equilibrium entrants), or SSS(EE), if for all $f \neq f^*$, $f \in BR(f^*)$ implies that for all $\bar{\varepsilon} > 0$, there exist $\varepsilon \in (0, \bar{\varepsilon})$, $i \in I$, and $g_i \in BR_i(f^*)$ such that

$$\Pi_i((1 - \varepsilon)f^* + \varepsilon f; g_i) > \Pi_i((1 - \varepsilon)f^* + \varepsilon f; f_i)$$

holds.

In the definition, the term "equilibrium entrants," or "EE," stems from the assumption that the strategies taken by the entrants constitute an equilibrium in the game restricted to the set of best responses to the original distribution. In particular, if the game is a two-person matching situation characterized by either (8.1) or (8.2), the definition is equivalent to stating that f^* is an SSS(EE) if for all $f \neq f^*, f \in BR(f^*)$ implies that there exist $i \in I$ and $g_i \in BR_i(f^*)$ such that $\Pi_i(f; g_i) > \Pi_i(f; f_i)$ holds. Indeed, in a two-person matching situation, the linearity of the payoff structure implies

$$\Pi_i((1 - \varepsilon)f^* + \varepsilon f; g_i) = (1 - \varepsilon)\Pi_i(f^*; g_i) + \varepsilon\Pi_i(f; g_i). \quad (8.5)$$

However, if we consider a three-person matching situation, the above is not equivalent to the definition since (8.5) does not hold in general. The first property, which is stated without proof, is that any SSS(EE) distribution is a Nash equilibrium.

Proposition 8.5 *If a strategy distribution f^* is socially stable (against equilibrium entrants), then f^* is a Nash equilibrium.*

Next, we state the following main result, which says that the two definitions of social stability given above are equivalent.

Theorem 8.6 *A strategy distribution f^* is socially stable (with respect to the best response dynamics) if and only if it is socially stable (against equilibrium entrants).*

We need the following lemma to prove the theorem and some other subsequent results.

Lemma 8.7 *If there exist $f \in BR(f^*)$ and $\bar{\varepsilon} > 0$ such that for any $\varepsilon \in (0, \bar{\varepsilon})$, we have*

$$\Pi_i((1 - \varepsilon)f^* + \varepsilon f; f_i) \geq \Pi_i((1 - \varepsilon)f^* + \varepsilon f; g_i) \quad (8.6)$$

holds for all $i \in I$ and all $g_i \in BR_i(f^*)$, then there exists $T > 0$ such that the path p with $p(t) = (1 - e^{-t})f + e^{-t}f^*$, $t \in [0, T]$, is a best response path.

Proof Suppose there exist $f \in BR(f^*)$ and $\bar{\varepsilon} > 0$ that satisfy (8.6) for all $\varepsilon \in (0, \bar{\varepsilon})$. By the upper hemicontinuity of $BR(\cdot)$, there exists $T > 0$ such that for all $t \in [0, T]$,

$$\Pi_i((1 - e^{-t})f + e^{-t}f^*; f_i) \geq \Pi_i((1 - e^{-t})f + e^{-t}f^*; g_i)$$

holds for all $i \in I$ and all $g_i \notin BR_i(f^*)$. Therefore, the path p with $p(t) = (1 - e^{-t})f + e^{-t}f^*$ ($t \in [0, T]$) is a best response path departing from f^*. $\qquad\square$

Now, the proof of Theorem 8.6 is presented below.
Proof of Theorem 8.6 First, suppose that $f^* \in \mathcal{F}$ is not an SSS(EE). Then by the definition, there exist $f \in BR(f^*)$ and $\bar{\varepsilon} > 0$ such that for all $\varepsilon \in (0, \bar{\varepsilon})$, (8.6) of Lemma 8.7 holds. By the lemma, there exists a best response path departing from f^*.
Second, suppose that f^* is an SSS(EE). Suppose the contrary, i.e., that there exists a best response dynamic path departing from f^*. Assume without loss of generality

that $t = 0$ is a point of departure, namely, $p(0) = f^*$ and $p(\Delta t) \neq f^*$ for sufficiently small $\Delta t > 0$. Since b is a step function as well as right continuous, and since $b(\Delta t) \neq f^*$ has to hold for a sufficiently small $\Delta t > 0$, there exists $\bar{t} > 0$ such that for $t \in [0, \bar{t})$, $b(t) = f \neq f^*$ holds. By the definition of SSS(EE), there exist $\bar{\varepsilon} > 0$, $i \in I$, and $g_i \in BR_i(f^*)$ such that

$$\Pi_i((1 - \varepsilon)f^* + \varepsilon f; f_i) < \Pi_i((1 - \varepsilon)f^* + \varepsilon f; g_i)$$

holds for all $\varepsilon \in (0, \bar{\varepsilon})$. In addition, we have

$$\Pi_i(f^*; f_i) = \Pi_i(f^*; g_i).$$

Thus, for all $t \in (0, \bar{t})$,
$$f_i \notin BR_i(p(t))$$

holds, which is a contradiction. □

Since we have established the equivalence result, we will refer to socially stable strategies with no qualification.

8.3.5 Set-Valued Solution Concepts and Their Existence

We have found a static expression corresponding to the stable points of a dynamic process. Throughout the previous sections, we did not pay attention to the existence of a solution. SSS may not exist. Addition of redundant strategies leads to nonexistence. For a game, not having any socially stable strategies simply means that there are no stable points with respect to the best response dynamics. Unlike existence in the context of strategic stability, nonexistence is not a shortcoming of the concept, but is rather a natural consequence of the actual dynamics. However, since we are considering the actual dynamic process occurring in real time, we would like to pin down the set of distributions, if not a singleton, into which the behavior pattern is absorbed. More precisely, we would like to include two or more distinct strategy distributions in the same "stable" set of points if there are best response paths to and from each other.

To express this idea, we have the following definition of accessibility. First, using best response paths, we say that a strategy distribution g is *directly accessible* from f if there exists a best response path $p : [0, T] \to \mathcal{F}$ for some $T > 0$ such that $p(0) = f$ and $p(T) = g$. The notion of accessibility is then recursively defined. We say that y is accessible from f if at least one of the following is satisfied: (i) g is directly accessible from f; (ii) there exists a sequence (g_n) converging to g such that g_n is accessible from f for all $n = 1, 2, \ldots$; and (iii) g is accessible from h which in turn is accessible from f. Now, our solution concept is as follows.

Definition 8.12 A nonempty subset F^* of $\mathcal{F}$ is called a socially stable set (with respect to the best response dynamics), or SS set(BR), if
  (i) any $g \notin F^*$ is not accessible from any f in F^*; and
  (ii) every $g \in F^*$ is accessible from every f in F^*.

An SS set(BR) is stable in the sense that once the actual behavior pattern falls in the set, another strategy distribution may be realized if and only if it is contained in the SS set(BR).

Before giving the static equivalent of the above concept, we define feasible direction. Given $F \subset \mathcal{F}$ and $f \in F$, a vector $w \in \mathbb{R}^{\sum_{i \in I} |S_i|}$ is a *feasible direction* from f in F if there exists $\bar{\varepsilon} > 0$ such that $f + \varepsilon w \in F$ for all $\varepsilon \in [0, \bar{\varepsilon}]$.

Definition 8.13 A subset F^* of $\mathcal{F}$ is called a evolutionarily stable set (against equilibrium entrants), or ES set(EE) if it is minimal among those sets having the following properties:
  (C) F^* is nonempty and closed; and
       for each $f \in F^*$, $w \in \mathbb{R}^{\sum_{i \in I} |S_i|}$ is a feasible direction from $f \in F^*$
       whenever $g = f + w$ is in $\mathcal{F}$, and there exists $\bar{\varepsilon} > 0$ such that for all $\varepsilon \in [0, \bar{\varepsilon}]$,
           $\Pi_i(f + \varepsilon w; g_i) \geq \Pi_i(f + \varepsilon w; h_i)$
       holds for all $i \in I$ and all $h_i \in BR_i(f)$.

Now we state the equivalence.

Theorem 8.8 *A nonempty subset F^* is a socially stable set (with respect to the best response dynamics) if and only if it is an evolutionarily stable set (against equilibrium entrants).*

The proof of the equivalence result requires several steps, which we divide into some lemmata. First of all, for each $f \in \mathcal{F}$, we define $R(f)$ to be

$$R(f) = \{g \in \mathcal{F} \mid g \text{ is accessible from } f\}.$$

It is easy to verify that $R(f)$ is nonempty and closed for all $f \in \mathcal{F}$. Let $\subset$ stand for the normal inclusion relation, Then $(\{R(f)\}_{f \in \mathcal{F}}, \subset)$ is a partially ordered set. We have the following important lemmata, which are slight modifications of some steps of the proofs of the existence theorem (Theorem 8.3), and therefore, we relegate the proof to the existence proof of Sect. 8.2.

Lemma 8.9 $(\{R(f)\}_{f \in \mathcal{F}}, \subset)$ *has a minimal element.*

Lemma 8.10 $F^* \subset \mathcal{F}$ *is an SS set(BR) if and only if F^* is a minimal element in* $(\{R(f)\}_{f \in \mathcal{F}}, \subset)$.

A direct corollary of these lemmata is that at least one SS set(BR) always exists.

Theorem 8.11 (Gilboa and Matsui 1991) *For any G, there exists at least one SS set(BR).*

Another important consequence is that for any $f \in \mathcal{F}$, there exists at least one SS set(BR) which is accessible from f. Note, however, that not all SS sets(BR) are attractors of the best response dynamics.

The next two lemmata relate $R(\cdot)$ to the alternative definition of stability, namely, ES set(EE).

Lemma 8.12 *For any $f \in \mathcal{F}$, $R(f)$ satisfies* (C).

Proof Take any $f \in \mathcal{F}$ and any $f' \in R(f)$. Suppose that there exist $g \in Br(f')$ and $\bar{\varepsilon} > 0$ such that for all $\varepsilon \in [0, \bar{\varepsilon}]$,

$$\Pi_i((1 - \varepsilon)f' + \varepsilon g; g_i) \geq \Pi_i((1 - \varepsilon)f' + \varepsilon g; h_i)$$

holds for all $i \in I$ and all $h_i \in Br_i(f')$. Then by Lemma 8.7, there exist $T > 0$ and a best response path $p : [0, T] \to \mathcal{F}$ with $p(t) = e^{-t}f' + (1 - e^{-t})g$. Therefore, $g - f'$ is a feasible direction from f' in $R(f)$. Thus, $R(f)$ satisfies (C). $\square$

Lemma 8.13 *Suppose that $F \subset \mathcal{F}$ satisfies* (C). *Then there exists $f \in \mathcal{F}$ such that $R(f) \subset F$.*

Proof Take any $f \in \mathcal{F}$. Consider $R(f)$. Since f is contained in both $R(f)$ and F, $R(f) \cap F$ is nonempty. Suppose that $R(f)$ is not a subset of F. Then there exists a path departing from some $f' \in R(f) \cap F$ into $R(f) \setminus F$. Suppose without loss of generality that f' is the point of departure into $R(f) \setminus F$. Let $p : [0, T] \to \mathcal{F}$ be such a path for some $T > 0$ where $p(\Delta t) \in R(f) \setminus F$ for a sufficiently small $\Delta t > 0$. Then it must be the case that there exists $\bar{t} > 0$ such that $w = b(t) - f'$ is not a feasible direction from $f' \in F$ for all $t \in [0, \bar{t})$ since b is a right continuous step function. Then by (C), w cannot be a best response to $f' + \varepsilon w$ for any sufficiently small $\varepsilon > 0$, which is a contradiction. $\square$

Summarizing these lemmata, we have the proof of Theorem 8.8.

Proof of Theorem 8.8 Let $\mathcal{H}$ be the class of the subsets of $\mathcal{F}$ that satisfy (C). First, if H is a SS set(BR), then H is minimal in $\mathcal{H}$ by virtue of previous lemmata. Hence, H is a ES set(EE).

Second, suppose that H is an ES set(EE), which means that it is a minimal element of $\mathcal{H}$ with respect to $\subset$. Then by Lemma 8.13, there exists $R(f)$ for $f \in \mathcal{F}$ such that $R(f) \subset H$ holds. By Lemma 8.12, $\mathcal{H}$ contains $\{R(f)\}_{f \in \mathcal{F}}$. Therefore, H is also a minimal element of $\{R(f)\}_{f \in \mathcal{F}}$. Hence, Lemma 8.10 implies that H is a SS set(BR). $\square$

The equivalence result allows us to call both SS set(BR) and ES set(EE) stable sets socially stable set without any qualifications.

8.3.6   Various Games

Situations analyzed in game theory have become so diverse that no single solution concept can capture the human intuition in a general manner. When we try to analyze

Fig. 8.2 Forward induction

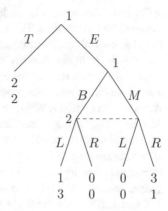

a new game, we often create a new concept. We then come up with another example in which this concept cannot capture the intuition. This so-called rat race seems endless if we view "sophisticated" rationality as something that does not change in the real world. Moreover, equilibrium theory is valid only when players are playing an equilibrium. It does not describe what happens if players fail to coordinate their behavior. The present analysis assumes the minimal amount of rationality which has been required in game theory, namely, people tend to take strategies that are best responses to the current situation regardless of whether it is in a socially stable set or not.

Yet, our result is powerful enough to capture our intuition on forward induction in the game shown in Fig. 8.2 (Kohlberg and Mertens 1986). In this game, there are two sequential equilibrium payoffs. One is (2, 2), and the other is (3, 1). Kohlberg and Mertens (1986) argues that (2, 2) is eliminated from the set of equilibrium outcomes in the following way. Player 1 (he) chooses T if player 2 (she) takes L with the probability of more than $1/3$. However, player 1 can make the following speech and take E followed by M (Kohlberg and Mertens 1986):

> Look, I had the opportunity to get 2 for sure, and nevertheless I decided to play in this subgame, and my move is already made. And we both know that you can no longer talk to me, because we are in the game, and my move is made. So think now well, and make your decision.

Listening to this speech, player 2 realizes that player 1 took T because he expects at least 2, and that his expectation is fulfilled only when she takes R with the probability of more than $1/3$, in which case he prefers M to B. Thus, taking E implies that he is going to take M for sure. If player 2 accepts the argument, then she will take R. But, does she necessarily have to be persuaded? The answer seems to be in the affirmative if it is common knowledge that both players are rational. In the real world, however, player 2 may stick to L by saying before the entire game begins:

> I suspect there is a possibility that you will make a mistake by choosing E. With this belief, if E is observed, I will conclude that you made a mistake and I will take L since you are supposed to take B. After listening to my story, you had better take T. It is still up to you. But remember my reasoning.

After this argument, it is unclear if $(T, B; L)$ should be eliminated from the set of "reasonable" equilibria. The argument on forward induction in this example works better in evolutionary settings. If the same situation is repeated many times in a large society, the behavior pattern may gradually change. Those type 1 individuals who attempt to experiment by taking E may take M. If the number of individuals taking EM is much greater than those taking EB, type 2 individuals gradually respond to the situation by taking R. Once a sufficient number of type 2 people convert to taking R, type 1 individuals start taking EM

This idea is captured by the socially stable set, and the unique socially stable set is the singleton $\{([EM], [R])\}$, although it is a concept in normal-form games and hence we have to look at its reduced normal form, as shown in Table 8.5. In this game, since $([EM], [R])$ is a strict Nash equilibrium, it is a socially stable strategy. Moreover, if you start with $([T], [L])$, there exists a best response path $p : [0, \bar{t}] \to \mathcal{F}$ $(\bar{t} > \ln 3)$ expressed as

$$
p(t) = \begin{cases} (1 - e^{-t})([T], [R]) + e^{-t}([T], [L]) & \text{if } t \in [0, \ln 3], \\ (1 - e^{-t})([T], \frac{1}{3}[L] + \frac{2}{3}[R]) + e^{-t}([EM], [R]) & \text{if } t \in (\ln 3, \bar{t}]. \end{cases}
$$

Therefore, $([T], [L])$ cannot be a member of any socially stable set; starting from any other point, it is verified that $([EM], [R])$ is accessible from the point. Hence, $([EM], [R])$ is the unique socially stable set in this example.

The present analysis proposes ESS-like static concepts that correspond to stable points of a dynamic process called best response dynamics. A point-valued stability concept is called a socially stable strategy, i.e., a strategy distribution that has no "robust" mutant. The concept is weaker than ESS since in order to upset a socially stable strategy, a mutant must fare not only better than the original distribution but also better than any other mutant. A socially stable strategy may not exist. Non-existence of a stable point does not bother us; rather, it is a natural consequence of the dynamic process. We then consider a set-valued concept called a socially stable set. A socially stable set is a minimal set with respect to the property that no "robust" mutant drives the behavior pattern away from the set. We showed its existence. The equivalence results between static and dynamic concepts of stability frees us to use those ESS-like static concepts without worrying about their dynamic justification. Our result shows that the concept captures our intuition on forward induction at least in some examples.

Table 8.5 The reduced normal form

	L	R
T	2, 2	2, 2
EM	0, 0	3, 1
EB	1, 3	0, 0

8.4 Deterministic versus Stochastic Dynamics

This book uses deterministic dynamics rather than stochastic dynamics in the analysis. We need to explain why this is the case. If one's motivation is to select equilibrium irrespective of the initial condition, deterministic dynamics is not a real contender since every strict Nash equilibrium is a stable point of any deterministic dynamics. Foster and Young (1990) and Kandori et al. (1993) (henceforth, KMR) show that models with perpetual randomness can give very different results. Other related papers include Young (1993) and Fudenberg and Harris (1992). This section discusses KMR since their model is followed by papers addressing some important aspects in the study of cultural diversity.

Suppose that time is discrete, the horizon is infinite, and that there are N players who are randomly matched to play the game of common interest shown in Table 8.6. Note again that (L, L) is Pareto efficient, while (R, R) is risk dominant. Assume for simplicity that N is not divisible by three. Let $z_t \in \{0, 1, \ldots, N\}$ be the number of players who play R in the tth period. KMR considers a Markovian dynamics where z_t is used as a state. That is, the current number of players playing R solely determines the future. In a static situation, $z_t = 0$ and $z_t = N$ are strict Nash equilibria. Also, if $z < N/3$, then L is the unique best response, while if $z > N/3$, then R is the unique best response.

Like other work on evolution, KMR uses adaptation and mutation as sources of adjustment in this environment. Adaptation is expressed by a deterministic process $z_{t+1} = b(z_t)$. To capture the idea of adaptation, KMR makes the following assumption:

$$b(z) < z \text{ if } 0 < z < \frac{1}{3}N, \quad b(z) > z \text{ if } \frac{1}{3}N < z < N, \quad b(0) = 0, \quad b(N) = N.$$

On top of this, KMR also uses a process of mutation. To be precise, after each selection phase, each player independently switches her action with probability $\varepsilon > 0$, where ε is considered to be small. Due to the stationarity of these processes, a Markov chain is represented by a transition matrix, $B(\varepsilon)$. Because of independent mutation, any state can be reached from any other state in a single period with positive probability. This implies that there exists a unique stationary distribution. Let $\mu(\varepsilon)$ be the stationary distribution. KMR characterizes the limit of $\mu(\varepsilon)$ as ε goes to zero. They show that $\mu^* = \lim_{\varepsilon \to 0} \mu(\varepsilon)$ puts probability one on the risk dominant equilibrium, which is called a long run equilibrium (KMR) or stochastically stable (Foster and Young).

Table 8.6 The game of common interest where the Pareto efficient outcome is different from the risk dominant outcome

	L	R
L	4, 4	0, 3
R	3, 0	2, 2

The limit distribution μ^* tells us about the behavior of the system for small but positive ε. That is, for small $\varepsilon > 0$ the behavior pattern of the system stays at the risk dominant equilibrium most of the time. Since, however, the Markov chain is irreducible, the risk dominant equilibrium is not an absorbing state. Indeed, it is upset and any outcome is reached with probability one. Nothing lasts forever. Still, as ε approaches zero, the expected duration of stay in each equilibrium tends to infinity. Moreover, the expected length of stay in the risk dominant equilibrium relative to that in the risk dominated one goes to infinity as ε goes to zero.

The crude logic behind this is the following. Suppose that the current behavior pattern is the risk dominant equilibrium. If the number of mutants in a certain period is less than $2N/3$, then the deterministic dynamics brings the system back to the original equilibrium with high probability. It is only when the number of mutants exceeds $2N/3$ that the deterministic dynamics moves the system away from it. In this case, the system goes to the risk dominated equilibrium with high probability. On the other hand, only more than $N/3$ mutants are needed to go to the risk dominant equilibrium from the risk dominated equilibrium. The probability of having $N/3$ mutants in a single period is the order of $\varepsilon^{N/3}$. While the probability of having $2N/3$ mutants is the order of $\varepsilon^{2N/3}$. Thus, the ratio between the probability of the event in which the risk dominant equilibrium is "upset" and the probability of the event in which the risk dominated one is "upset", $\varepsilon^{N/3}$, tends to zero as ε goes to zero. That is, the risk dominant equilibrium is much less likely to be upset than the risk dominated equilibrium. Kandori and Rob (1995) extends this analysis to a broader class of games with more than two actions.

What the stochastic models have accomplished is to overturn the general feeling that if selection operates at a sufficiently higher rate than the rate of mutation, it is reasonable to focus on the dynamic implications of a single mutation as in the replicator dynamics.

This has a significant implication for economic theory in general since many studies rely on the notion of the law of large numbers, and stochastic models basically say that a large but finite population behaves quite differently from an infinite population provided that we consider only independent and identically distributed (i.i.d.) random shocks.

One may wonder what happens to the law of large numbers. When we study probability theory, we learn the law of large numbers and the central limit theorem. Consider an i.i.d. process $\{X_i\}_{i=1}^{\infty}$ with $E(X_1) = \mu$ and $var(X_1) = \sigma^2$. Let $S_n = \sum_{i=1}^{n} X_n$. Then the law of large numbers tells us that S_n/n converges to μ with probability one, i.e., $Pr(\lim_{n\to\infty} S_n/n = \mu) = 1$. The central limit theorem focuses on the behavior of $S_n/\sqrt{n}$. We know that it follows a normal distribution, i.e., $S_n/\sqrt{n} \sim N(\mu, \sigma^2)$. Yet, there is a third result called the large deviation principle that focuses on the tail behavior of S_n/n. This principle examines the behavior of S_n/n away from μ. We know

$$\frac{1}{n} \ln Pr[\frac{S_n}{n} \geq \mu + \alpha] \ \to \ -\sup_{\lambda \in \mathbb{R}}[\lambda\alpha - \ln E[e^{\lambda X_1}]].$$

This principle basically tells us how likely it is that an unlikely event occurs. This principle plays a central role when we consider stochastic mutations in a finite population as well as in an infinite population.

A question arises as to whether a deterministic infinite population model or a stochastic finite population model or a stochastic finite population model approximates reality better. It has been argued that an alternative to an infinite population is a finite but large population, studying its asymptotic behavior. Many authors seem to believe that this would be the "right" thing to do, and only appallingly complicated mathematical computation, if anything, should allow us to use infinite, simpler models. Gilboa and Matsui (1992) proposed a different viewpoint, according to which models with an infinite population are sometimes conceptually better than models with a finite population. Since the world consists of finitely many agents, a finite population model is more accurate objectively. However, being boundedly rational, each individual in a large population behaves as if he were too small to affect the society in any way. If our purpose is to analyze the behavior of such agents and its consequences, it may be better to construct an infinite population model reflecting the way the agents see the world.

In models of evolution, the same concern arises. If the size of the population is very large, it takes a very long time to flip from one equilibrium to another in, say, KMR, and it may not be the dominant force for evolution, at least in the time span we consider. After all, as Keynes put it, in the long-run, everyone will die. For these cases, deterministic models may often have better explanatory power. If we analyze the real world, we have to keep this point in mind and carefully choose between the two types of models.

Chapter 9
Cheap-Talk and Cooperation in a Society

9.1 Introduction

Does cheap-talk matter? Does costless communication necessarily yield Pareto optimal outcomes? When is it that the voices of those with disabilities reach the majority,
or persons without disability? These questions are relevant in the context of economy
and disability.[1] In many situations, persons without disability do not see the barriers
that persons with disability are faced with. If the barriers were removed simply by
pointing them out, life would be easier for persons with disability than now.

The answers to the above questions depend, of course, on the specific way of
modeling the interaction, and, most importantly, on the solution concept applied.
This section uses a solution concept adopting a societal-dynamics perspective, which
yields results that are quite different from those from more "traditional" methods.
Let us first briefly review some of the existing concepts.

In the field of noncooperative game theory, every solution concept requires that
players make rational choices based on their beliefs, and that those beliefs satisfy a
certain consistency. The criteria of rational choices and consistency of beliefs vary
from one concept to another. For example, in Nash equilibrium (Nash 1951), each
player maximizes her own (unconditional) expected payoff, given the other players'
strategies. Sequential equilibrium (Kreps and Wilson 1982) requires, in addition to
those conditions required in Nash equilibrium, that each player chooses an action
that maximizes her conditional expected payoff at every information set, and that the
conditional belief system is consistent with the equilibrium strategy profile.

Regarding normal form games, every equilibrium concept requires that players'
beliefs about the other players' strategies must coincide with the "actual" strategies
that the players plan to take. On the other hand, rationalizability (Bernheim 1984;
Pearce 1984) allows two players' beliefs to differ from the actual strategies as long
as they are derived in a rational manner.

---

[1]This chapter is based on Matsui (1991).

© Springer Nature Singapore Pte Ltd. 2019
A. Matsui, *Economy and Disability*, Economy and Social Inclusion,
https://doi.org/10.1007/978-981-13-7623-8_9

In spite of such a wide spectrum, there is a common assumption underlying these concepts, namely, that players' systems of beliefs and strategies do not change throughout the entire game. The game is played exactly once (if it is a repeated game, the repetition occurs once), and if they change their beliefs or strategies, the changes are incorporated in larger systems of beliefs and strategies. In this sense, each player is treated as if he had a complete contingent plan of beliefs and strategies. The stability of strategies discussed in this context is called strategic stability.

The analysis of cheap-talk models in this context raises some difficulties. Consider the game in Table 9.1. Although cooperation, namely the play (L, L), has been prescribed as a unique outcome under unlimited communication (see, for instance, Bernheim et al. 1987), no purely noncooperative solution concept discussed in the literature has successfully excluded (R, R) as an outcome. Indeed, even with cheap-talk, it is doubtful that (L, L) is always chosen when (R, R) is oriented. The reasoning for this situation is the following. If, say, player 1 wants to make a joint deviation to (L, L), he must persuade player 2 that he plans to take L with the probability of at least a half. Even if he is not sure that she is persuaded, he always gains by the persuasion if he sticks to R. Knowing this, she might conclude that it is safe to keep playing R. Hence, cheap-talk is not credible. On the other hand, if it is common knowledge that player 2 thinks the persuasion credible, then player 1 is willing to persuade her and take L. Both of these scenarios make some sense. Therefore, it seems that one must arbitrarily assume a certain degree of credibility of cheap-talk in refining equilibrium. We cannot escape from this type of arbitrariness if we try to provide cheap-talk with some meaning in the context of strategic stability. We will see that if one adopts a societal-dynamics perspective, a high degree of credibility may be attained, rather than assumed, in the course of a movement of a behavior pattern.

In many daily life situations, people do not know and/or do not care about the entire structure of the game. Nevertheless, they behave so as to maximize their payoffs. In order to behave optimally, all one has to know are one's own payoff and the opponents' strategies, or, to be precise, the expected payoff from each action available.

One of various plausible scenarios explaining how they learn to behave optimally is that the game is repeated many times, and people determine their actions by trial and error.' In this process, since the behavior pattern typically changes as time goes on, the belief system changes as well. Social stability refers to the stability of a stationary point in this repeated situation.

Table 9.1 Game of common interest

	L	R
L	4, 4	0, 3
R	3, 0	2, 2

Chapter 8 introduced a solution concept called a socially stable set on the basis of this way of viewing the world. This solution concept is applied to societies with infinitely many individuals who are matched randomly to play a single game. With repetition over a long time, the behavior pattern of society changes gradually in the direction of a best response. Its trajectory is referred to as a directly accessible path. A strategy distribution is accessible from another if either (i) there is a directly accessible path from the latter to any neighborhood of the former, or (ii) the former is accessible from another that is accessible from the latter. A socially stable set is a set of strategy distributions that is closed under accessibility, and any two members of which are accessible from each other. An interpretation is that once the actual behavior pattern falls in the set, it never leaves the set, and any strategy distribution in the set may always be realized.

In the present analysis, we apply this concept to the class of games with cheap-talk. In the main part of the analysis, we confine our attention to two-by-two games of common interest. We say that a game is of common interest if there exists a unique weakly (a fortiori a unique strongly) Pareto optimal outcome.

Then we consider the following two-person two-stage game. In the first stage, players simultaneously announce actions. In the second stage, they choose their actions after observing the first-stage announcements. Payoffs are determined only by the actions they take in the second stage. This two-stage game is called a "game with cheap-talk" since the actions taken in the first stage directly affect neither the payoffs nor the set of actions available to players in the second stage of the game. The second stage of the original game is referred to as the game without cheap talk.

In the context of social stability, we assume that this two-stage game itself is repeated in a large society with random matching. As time goes on, a norm endowing cheap-talk with a certain meaning may develop. In particular, such a norm may lead to a Pareto optimal outcome. Indeed, in the example of Table 9.1, cooperation is bound to emerge. While this fact is formally proven in the subsequent sections, its logic may be informally explained as follows.

Suppose that the initial behavior pattern is such that all the players announce R in the first stage and take R in the second stage no matter what is announced in the first stage. Note that this strategy is a best response to itself. Then, some people start trying a new strategy that prescribes L in the first stage, L in the second stage if both have announced L, and R otherwise. By adopting this strategy, one gains three if he meets a person with the same strategy and gains one if he meets a person taking the original strategy. This strategy fares better than the original strategy as well as the "cheating" strategy, according to which one announces L and takes R in the second stage no matter what has been announced. Once it prevails, taking R in the second stage can never be prescribed by a best response strategy (except in unreached information sets).

This verbal description essentially corresponds to a formal argument in the subsequent sections. Its logic involves two important points concerning social stability as distinguished from strategic stability. First, we deal with situations in which people continue learning their opponents' strategies on the basis of what happened in the past as well as in the current world. Through this learning process, the role of

cheap-talk gradually evolves. Second, we are interested in a society that consists
of many individuals, who are matched randomly to play the game with cheap-talk
and are never matched again in the future so that they are not involved in strategic
interaction such as punishment beyond a single matching. If the game is repeated
many times between the same individuals, then it is unavoidable for the analysis to
take into account the strategic interaction between different two-stage games, and
consequently the game should be considered as a (possibly infinitely) repeated one.
In that case, some studies have shown that the optimal outcome is necessarily attained
under some qualifications; among those studies are Aumann and Sorin (1989) and
Matsui (1989). The point of the present approach is that these long-term strategic
considerations are not needed to explain cooperation.

The approach presented here is also different from the evolutionarily stable strategy proposed by Maynard Smith and Price (1973). The main distinction is that their
solution concept is point-valued, while ours is set-valued. A set-valued solution concept is more natural than a point-valued one in the context of social or evolutionary
stability since it captures not only the stability of a point, but also the stability of a
cyclical movement, if any. It turns out that our solution always exists, and that we
always find a socially stable set that is accessible from any initial distribution.

9.2   Games of Common Interest with Cheap-Talk

In a society, there are two types of individuals, 1 and 2; each type consists of infinitely
many anonymous individuals. They are matched randomly to take some actions. In
each matching situation, one individual from each type is selected, and they play
a two-stage game. In the first stage of each game, type 1 and type 2 players simultaneously announce either L or R. In the second stage, knowing what has been
announced, they again simultaneously take either L or R. Therefore, each individual
has five information sets, $u_i^0, u_i^1, u_i^2, u_i^3$, and u_i^4 where u_i^0 is the information set in the
first stage, and u_i^1, u_i^2, u_i^3, and u_i^4 are the information sets that are reached when the
first stage announcements were (L, L), (L, R), (R, L), and (R, R), respectively (the
first and the second coordinates of each pair correspond to the announcements of
type 1 and type 2 players, respectively). A pure strategy of type i player ($i = 1, 2$),
$s_i = (s_i^0, s_i^1, s_i^2, s_i^3, s_i^4)$, is an element of $S_i = \{L, R\}^5$ where s_i^k ($k = 0, 1, \ldots, 4$) is
the action taken at the information set of u_i^k. We write $S = S_1 \times S_2$. We assume
that a strategy distribution of type i ($i = 1, 2$) is a behavior strategy, which is expressed as a quintuple $f_i = (f_i^0, f_i^1, f_i^2, f_i^3, f_i^4)$ in $[0, 1]^5$ where f_i^k ($k = 0, \ldots, 4$) is
a "local" strategy at u_i^k, which prescribes L with probability f_i^k. Let $\mathcal{F}_i$ be the set of
all strategy distributions of type i. We write $\mathcal{F} = \mathcal{F}_1 \times \mathcal{F}_2$. We assume that $\mathcal{F}$ is a
subset of a 10-dimensional space endowed with the usual linear operations and the
norm $\| \cdot \|$ defined by $\|f\| = \sup_{i,k} |f_i^k|$. Given f in $\mathcal{F}$ and a subset G of $\mathcal{F}$, we let
$\|f - G\| = \inf_{g \in G} \|f - g\|$.

Given f in $\mathcal{F}$, there are two possible scenarios concerning choices of strategies
by individuals. One is that the f_i^0-fraction of the entire population of type i ($i = 1, 2$)

takes the pure local strategy L at u_i^0, and the f_i^k-fraction takes L conditional on u_i^k being reached. The other scenario is that every type i individual takes the same behavior strategy f_i. This distinction might not affect our analysis in the sequel. However, we find the former more appealing than the latter and prefer to keep it in mind. In considering the dynamic adjustment process, the current strategy distribution will be often referred to as the behavior pattern.

Given a pair of local strategies $f^k = (f_1^k, f_2^k)$ at information sets u_1^k and u_2^k ($k \neq 0$), respectively, the expected payoff for type i player given u_i^k being reached ($i = 1, 2$) is calculated as

$$\pi_i(f^k) = f_i^k f_j^k a_i + f_i^k (1 - f_j^k) b_i + (1 - f_i^k) f_j^k c_i + (1 - f_i^k)(1 - f_j^k) d_i,$$

where j denotes not i ($j \neq i$) hereafter.

The second stage of the game is given in Table 9.2. A game is said to be a game of common interest if there is a unique weakly Pareto optimal outcome. In the present analysis, we assume without loss of generality that a_i is strictly greater than b_i, c_i, and d_i for $i = 1, 2$. Let

$$\bar{q}_i = \frac{d_i - b_i}{a_i - c_i + d_i - b_i}, \quad i = 1, 2,$$

if the denominator is positive, and $\bar{q}_i = -\delta$ for some $\delta > 0$ otherwise. Then at u_i^k ($k \neq 0$, $i = 1, 2$), we have

$$\arg\max_{f_i^k \in [0,1]} \pi_i(f_i^k, f_j^k) = \begin{cases} = \{1\} & \text{if } f_j^k > \bar{q}_i, \\ \in [0, 1] & \text{if } f_j^k = \bar{q}_i, \\ = \{0\} & \text{if } f_j^k < \bar{q}_i. \end{cases}$$

Note that if $\bar{q}_i$ is nonpositive, it is always a best response for player i to take L in the second stage.

Given a strategy profile $f \in \mathcal{F}$, the expected payoff for an individual of type i ($i = 1, 2$) if he takes a behavior strategy $g_i \in \mathcal{F}_i$ is

$$\Pi_i(f; g_i) = g_i^0 f_j^0 \pi_i(g_i^1, f_j^1) + g_i^0(1 - f_j^0)\pi_i(g_i^{1+i}, f_j^{1+i})$$
$$+ (1 - g_i^0) f_j^0 \pi_i(g_i^{4-i}, f_j^{4-i}) + (1 - g_i^0)(1 - f_j^0)\pi_i(g_i^4, f_j^4).$$

We may write $\Pi_i(f) = \Pi_i(f; f_i)$ and $\Pi(f) = (\Pi_1(f), \Pi_2(f))$.

Table 9.2 A general 2-by-2 game of common interest

	L	R
L	a_1, a_2	b_1, c_2
R	c_1, b_2	d_1, d_2

Let $BR_i(f)$ be the set of strategies for individuals of type i ($i = 1, 2$) that are best responses to f, i.e.,

$$BR_i(f) = \arg\max_{g_i \in \mathcal{F}_i} \Pi_i(f; g_i).$$

We may write $BR(f) = BR_1(f) \times BR_2(f)$. Given $f \in \mathcal{F}$, let $U_\varepsilon(f)$ be the ε-neighborhood of f, i.e., $U_\varepsilon(f) = \{g \in \mathcal{F} \mid \|g - f\| < \varepsilon\}$.

9.3   Social Stability and Socially Stable Sets

This section defines and discusses the concept of stability. To capture the idea of social stability, we consider the following two points. First, there are no strategic considerations between any two matchings (although there may be strategic interaction within each matching situation), which reflects the assumption that individuals are anonymous and are matched randomly. Second, unlike a deviation made by a single player, a change in behavior pattern is likely to be continuous. This expresses the idea that within a small time interval, only a small portion of individuals change their strategies. In order to express these points, the notion of accessibility is given, which is similar to the one defined in Chap. 8.

Definition 9.1 A strategy distribution g is directly accessible from another strategy distribution f if there exist a continuous $p : [0, 1] \to \mathcal{F}$ that is differentiable from the right, $q : [0, 1) \to \mathcal{F}$ that is continuous from the right, and $\alpha > 0$ such that $p(0) = f$, $p(1) = g$,

$$\frac{d^+}{dt} p(t) = \alpha[q(t) - p(t)], \quad \text{for } t \in [0, 1),$$

and

$$q(t) \in BR(p(t)), \quad t \in [0, 1).$$

In the definition, $\frac{d^+}{dt} p(t)$ is the right derivative of p at t. The definition says that the behavior pattern $p(t)$ moves in the direction of a best response to itself (toward $q(t)$), and it may stay at the same place only when the behavior pattern is a best response to itself. The interpretation of this definition is that only small and equal fractions of individuals in each type recognize the current behavior pattern and change their strategies to those that are the best responses to it. We call the function p a directly accessible path from f to g. Then a strategy distribution g is said to be *accessible* from another distribution f if at least one of the following is satisfied:

  (i) g is directly accessible from f;
 (ii) there exists a sequence $(g_n)_{n=1}^\infty$ converging to g such that g_n ($n = 1, 2, \ldots$) is accessible from f;
(iii) g is accessible from some h that in turn is accessible from f.

Using this notion of accessibility, we are now in a position to present the definition of a socially stable set.

Definition 9.2 A nonempty subset F^* of $\mathcal{F}$ is a socially stable set if no $g \notin F^*$ is accessible from any $f \in F^*$, and every g in F^* is accessible from every f in F^*.

A socially stable set is stable in the sense that once the actual behavior pattern falls into it, another strategy distribution may be realized if and only if it is within the socially stable set. The interpretation of this concept is as follows: for a long time, individuals search for better strategies. After they sufficiently experience, a behavior pattern falls into a socially stable set, may move within it, but never leaves it.

9.4 Optimality Result and Its Proof

This subsection presents the main theorem of this section and its proof. Let $F^{**} \subset \mathcal{F}$ be defined as

$$F^{**} = \{f \in \mathcal{F} |\ \Pi(f) = (a_1, a_2)\}.$$

Note that F^{**} is closed, and that f is in F^{**} if and only if $f^k \neq (1, 1)$ $(k \neq 0)$ implies that u_i^k is not reached. Now, we are in a position to state the following result.

Theorem 9.1 F^{**} *is the unique socially stable set in any two-by-two game of common interest with cheap-talk.*

Note that the statement of the theorem does not hold in a game without cheap-talk. Consider the game of Table 9.1. If all the individuals take R, then none has an incentive to take L and get zero instead of one. Therefore, (R, R) is a socially stable strategy, i.e., it forms a socially stable set as a singleton.

On the other hand, in the game with cheap-talk, it is possible for individuals to cooperate and attain the optimal outcome because even if all the rest take R, one can take the strategy that expects a "signal" and takes L if one gets the signal and remains at R if one does not. This is followed by the opponent's change of the strategy to the one which actually sends a signal for cooperation.

Next, we present the proof of the theorem, which consists of some lemmata. The first lemma states that for any strategy distribution, there exists another strategy distribution accessible from it that prescribes either (L, L) or (R, R) at each information set of the second stage.

Lemma 9.2 *Given* $f \in \mathcal{F}$, *there exists* g *in* $\mathcal{F}$ *that is accessible from* f *such that for each* $k = 1, 2, 3, 4$, *either* $g^k = (0, 0)$ *or* $g^k = (1, 1)$ *holds.*

Proof Given $f \in \mathcal{F}$ and $\varepsilon > 0$, we first construct p and q satisfying the conditions of direct accessibility such that $p^k(1)$ is in the ε-neighborhood of $(0, 0)$ or $(1, 1)$ for all $k = 1, \ldots, 4$. As is seen from the expression of expected payoff, q^k $(k \neq 0)$ can be determined independently of $p^{k'}$ and $q^{k'}$ for $k' \neq k$. Let q_i^k ($k \neq 0, i = 1, 2$) be defined as

$$q_i^k(t) = \begin{cases} 1 & \text{if either } p_j^k(t) > \bar{q}_j \text{ or } [p_j^k(t) = \bar{q}_j \wedge p_i^k(t) > \bar{q}_i], \\ 0 & \text{otherwise.} \end{cases}$$

Then $q_i^k(t)$ maximizes $\pi_i(\cdot, p_j^k(t))$. By letting

$$\frac{d^+}{dt} p^k(t) = \alpha[q^k(t) - p^k(t)],$$

$p^k(1)$ approaches $(1, 1)$ or $(0, 0)$ as α becomes larger. Figure 9.1 illustrates the dynamics for the case of $d_i - b_i > 0$ for $i = 1, 2$.

Next, we consider the dynamics of $p^0(t)$. Let

$$L_i(p^0, t) = p_j^0 \pi_i(p_i^1(t), p_j^1(t)) + (1 - p_j^0)\pi_i(p_i^{1+i}(t), p_j^{1+i}(t)),$$
$$R_i(p^0, t) = p_j^0 \pi_i(p_i^{4-i}(t), p_j^{4-i}(t)) + (1 - p_j^0)\pi_i(p_i^4(t), p_j^4(t)).$$

Then let $B_i(p^0, t)$ ($i = 1, 2$) be given by

$$B_i(p^0, t) \begin{cases} = \{1\} & \text{if } L_i(p^0, t) > R_i(p^0, t), \\ \in [0, 1] & \text{if } L_i(p^0, t) = R_i(p^0, t), \\ = \{0\} & \text{if } L_i(p^0, t) < R_i(p^0, t). \end{cases}$$

The expression $L_i(p^0, t)$ is the expected payoff of type i player if he takes L at u_i^0, i.e., chooses $f_i^0 = 1$. Similarly, $R_i(p^0, t)$ is the expected payoff of type i player if he takes R at u_i^0, i.e., chooses $f_i^0 = 0$. We may write $B(p^0, t) = (B_1(p^0, t), B_2(p^0, t))$.

Fig. 9.1 Phase diagram for the second stage of the coordination game with cheap-talk

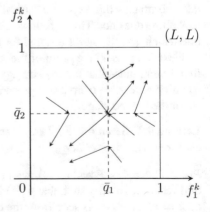

Consider the following problem of differential inclusion:

$$\frac{d^+}{dt}p^0(t) \in \alpha[B(p^0, t) - p^0(t)], \quad p^0(0) = f^0.$$

Note that any $q(t)$ with $q^0(t) \in B(p^0, t)$ and $q^k(t)$ defined as above is in $BR(p(t))$. Because of the existence theorem of differential inclusion (see, e.g., Aubin and Cellina 1984, p. 98), the above problem has a solution since $B(p^0, t)$ is upper hemicontinuous and convex-valued. Therefore, for each $\varepsilon > 0$, there exists a directly accessible path p such that $p^k(1) \in U_\varepsilon((0, 0)) \cup U_\varepsilon((1, 1))$ holds for each $k = 1, 2, 3, 4$.

Let (ε_n) and (g_n) be sequences such that (ε_n) converges to zero, and $g_n^k(1)$ is in $U_{\varepsilon_n}((0, 0)) \cup U_{\varepsilon_n}((1, 1))$. Then there exists a convergent subsequence (g_{n_ℓ}). Let g be the limit of (g_{n_ℓ}). Then g is accessible from f and $g^k = (0, 0)$ or $(1, 1)$ for each $k \neq 0$. $\qquad\square$

The next lemma states, by making use of Lemma 9.2, that for any strategy distribution, there exists another distribution in F^{**} which is accessible from it.

Lemma 9.3 *For any $g \in \mathcal{F}$, there exists $f \in F^{**}$ that is accessible from g.*

Proof From Lemma 9.2, it is sufficient to show that for any g such that for each $k \neq 0$, $g^k = (0, 0)$ or $(1, 1)$ holds, there exists $f \in F^{**}$ that is accessible from g. Consider the case in which $g^k = (0, 0)$ holds for all $k \neq 0$. The result is similarly proven for the other cases as well. Let h be such that $h^k = (0, 0)$ holds for $k = 0$ as well as $k \neq 0$. For any $\varepsilon > 0$, there exists $h^\varepsilon \in U_\varepsilon(h)$ that is directly accessible from g. Indeed, given $0 < \varepsilon < 1, p : [0, 1] \to \mathcal{F}$ given by

$$p^0(t) = \varepsilon^t g^0 + (1 - \varepsilon^t)h^0, \quad p^k(t) = (0, 0)$$

is a directly accessible path from g to $\varepsilon g + (1 - \varepsilon)h$ with $b(t) = h$ for all t and $\alpha = -\ln \varepsilon$. Thus, h is accessible from g.

Now, consider $f \in \mathcal{F}$ with $f^0 = f^1 = (1, 1)$ and $f^k = (0, 0)$ for $k = 2, 3, 4$. Note that f is in F^{**}. Then a path p defined as $p(t) = \varepsilon^t h + (1 - \varepsilon^t)f$ is a directly accessible path from h to $\varepsilon h + (1 - \varepsilon)f$ with $b(t) = f$ for all t and $\alpha = -\ln \varepsilon$. Indeed, $b(t) \in BR(p(t))$ holds for all t since we have $f \in BR(h)$ and $f \in BR(f)$ since we have $h^0 = (0, 0)$. Thus, a strategy distribution that attains (a_1, a_2) is accessible from any strategy distribution by virtue of Lemma 9.2 and the transitivity of accessibility. $\qquad\square$

Now, we present the proof of the theorem.

Proof of Theorem Since we establish the fact that for any strategy distribution g, there exists f in F^{**} that is accessible from g, what we have to prove is that any two strategy distributions in F^{**} are mutually accessible and that any g outside F^{**} is not accessible from any strategy distribution in F^{**}. Since F^{**} is connected, and pairs of payoffs of all the elements in F^{**} are identical, it is clear that any two members in F^{**} are mutually accessible.

Next, take any f in F^{**}. Suppose the contrary, i.e., that there exists $g \notin F^{**}$ that is accessible from f. Then there exist $p : [0, 1] \to \mathcal{F}$, $b : [0, 1) \to \mathcal{F}$, and $\alpha > 0$ that satisfy the definition of direct accessibility since F^{**} is closed. We may assume without loss of generality that f is a point of departure at time 0, i.e., for any small $\Delta t' > 0$, there exists $\Delta t < \Delta t'$ such that $p(\Delta t) \notin F^{**}$ holds. We divide the proof into three cases.

First, suppose that $0 < f_i^0 < 1$ holds for $i = 1, 2$. Then $f^k = (1, 1)$ holds for $k \neq 0$. Let ε satisfy $0 < \varepsilon < \frac{1}{2} \min\{f_1^0, 1 - f_1^0, f_2^0, 1 - f_2^0, 1 - \bar{q}_1, 1 - \bar{q}_2\}$. Take any h in $U_\varepsilon(f)$. Since $0 < h_i^0 < 1$ $(i = 1, 2)$ and $h^k > (1 - \bar{q}_1, 1 - \bar{q}_2)$ hold, $b \in BR(h)$ implies that $b^k = (1, 1)$ for $k \neq 0$. Therefore, $\|p(t) - F^{**}\|$ is decreasing in t. Thus, no path can leave F^{**} at f.

Second, suppose that $0 < f_i^0 < 1$ and $f_j^0 = 0$ or 1 hold. Assume without loss of generality that $0 < f_2^0 < 1$ and $f_1^0 = 0$ hold. Then since u_i^1 and u_i^2 $(i = 1, 2)$ are not reached, f^1 and f^2 may be anything, while $f^3 = f^4 = (1, 1)$ must hold. Again, for h in any sufficiently small neighborhood of f, $b \in BR(h)$ implies that $b_1^0 = 0$ whenever either h^1 or h^2 is different from $(1, 1)$, which means that $\|p(t) - F^{**}\|$ is decreasing in t.

Finally, suppose that f^0 is one of (0,0), (1,0), (0,1), and (1,1). Assume without loss of generality that $f_1^0 = f_2^0 = 0$ holds. Then $f^4 = (1, 1)$ holds. Take any h in a sufficiently small neighborhood of f, $b \in BR(h)$ implies that we have $b_1^0 = 0$ if $h^2 \neq (1, 1)$, $b_2^0 = 0$ if $h^3 \neq (1, 1)$, and $b^0 = (0, 0)$ if $h^2 = h^3 = (1, 1)$ and $h^1 \neq (1, 1)$ hold. This again implies that $\|p(t) - F^{**}\|$ is always decreasing in t. Thus, there is no directly accessible path from any f in F^{**} to g outside F^{**}. $\qquad\square$

9.5   Other Games

This section discusses some other games. First of all, the optimality result can be extended to the class of two-person (not necessarily two-by-two) pure-coordination games with cheap-talk in which the payoffs to both players are identical.

Next, we consider the game of the "battle of the sexes" shown in Table 9.3. It is shown that in this game, a pair of payoffs (1.5, 1.5) is attained in a socially stable set, namely that cheap-talk serves as a correlation device. Suppose that L is taken by type i players $(i = 1, 2)$ at u_i^1 and u_i^4 and R is taken at u_i^2 and u_i^3, i.e., f^k is (1, 1) for $k = 1, 4$ and (0,0) for $k = 2, 3$. Furthermore, suppose that $f_1^0 = f_2^0 = 1/2$ holds.

Table 9.3 The battle of the sexes

	L	R
L	2, 1	0, 0
R	0, 0	1, 2

As observed in the previous section, since all the information sets are reached with positive probability, and since each player strictly prefers to keep the present action at each information set of the second stage, we conclude that f_i^k ($k \neq 0, i = 1, 2$) is not changed in the neighborhood of f^0. Moreover, there is no directly accessible path for the behavior pattern to move away from f^0 if the f_i^k values ($k \neq 0, i = 1, 2$) do not change as can be seen in Fig. 9.2. Thus, $\{f\}$ is a stable set as a singleton, and $\Pi(f) = (1.5, 1.5)$.

It is not always the case that Pareto efficient outcomes are selected in the course of social adjustment (see Table 9.4). One may prove that (B, R) is in a socially stable set. Indeed, in this game, (B, L) is always accessible from (T, L), and (B, R) is also accessible from them. Therefore, a socially stable set consists of (B, L), (T, L), (B, R), and some convex combinations of them. This game has a unique strongly Pareto optimal outcome. Thus, the uniqueness of the weakly, not only strongly, Pareto optimal outcome is essential for the optimality result.

We have seen that socially stable sets predict the set of Pareto optimal outcomes to be the only solutions in two-by-two games of common interest with cheap-talk. The point of this analysis is not only that cooperation is necessarily achieved, but also that it is achieved without any presumption on the credibility of cheap-talk. A high degree of credibility of cheap-talk is attained, rather than assumed, in the societal dynamic process. In other words, "truth becoming focal" is in our analysis instead of "truth being focal" as in Farrell (1993) and Rabin (1990).

This result reinforces the validity of the solution concept when we consider random matching games in a large society, especially those games in which players' reasoning processes alone cannot sort out a single equilibrium.

Fig. 9.2 Phase diagram for the battle of the sexes with cheap-talk

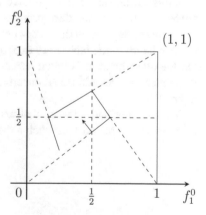

Table 9.4 A game with a unique strongly pareto optimal outcome

	L	R
T	2, 2	0, 0
B	2, 1	1, 1

9.6   Voice Matters in a Dictator Game

The analysis of this chapter as well as Crawford and Sobel (1982) indicates that cheap-talk matters in strategic settings. At the same time, it has been shown that a coordination aspect is necessary for the validity of cheap-talk. In many situations, if not all, the game encountered by a person with disability is similar to a dictator game, wherein the person with disability asks a favor to the person without disability.

Yamamori et al. (2008) conducted an experiment to study the effects of voice or cheap-talk in a dictator game, while maintaining subjects' anonymity. In the experiment, the recipient has an opportunity to state a payoff-irrelevant request for his/her share before the dictator dictates his/her offer. They found that the independence hypothesis that voice or cheap-talk does not matter is rejected. In particular, if the request is for less than half of the pie, the dictator's offer increases as the recipient's request increases. Additionally, there is no dictator who is other-regarding and, at the same time, does not react to the recipient's request.

Also, when a concerned party asks for a share of some resources controlled by a dominant party, the concerned party sometimes asks a disinterested third party to mediate between the two instead of directly asking the dominant party. Workers with a disability often hesitate to ask their manager to create a barrier-free environment as they are afraid that such requests may be interpreted as greedy requests.

Yamamori et al. (2010) conducted another experiment on a dictator game with the option for a "voice" by a third party and compared it with the dictator game with a "voice" by the recipient. Their findings are as follows. The dictators' offers in response to an aggressive voice of the recipients are significantly lower than the dictators' offers in response to the corresponding voice of the third party. The dictators' responses to an aggressive voice differentiate the effects of the recipient's voice from those of the third party's.

There is a significant difference between the laboratory experiment and the reality. On one hand, in the laboratory experiment, it is clear what is a fair division, 50-50. On the other hand, it is much less clear what request is perceived as greedy. This difference induces additional uncertainty when we apply this experimental result to the reality.

We have not had a theory that explains these results yet, and it remains for the future research to construct such a theory.

Chapter 10
Evolution and the Interaction of Conventions

10.1 Introduction

The present chapter and the next analyze the interaction of conventions, or different behavior patterns. This chapter presents a general framework to analyze the interaction, while Chap. 11 focuses more on minority-majority issues in an economic arena. This chapter develops a two-society model to analyze the interaction of societies with different conventions. The two societies may be interpreted as two groups within a single society such as persons with and without disability. Different groups have different traits/preferences so that optimal actions and desirable social outcomes may be different from each other.

Conventions are the modes of behavior to which the majority of a society subscribe.[1] People follow a certain convention because others do the same. People sometimes stop following a certain mode of behavior when others start behaving in a different manner. Conventions established in one region change over time, and these changes are sometimes caused by technological innovations and the development of new ideas. These changes sometimes occur within a region. In other cases, changes occur when societies with different societal backgrounds begin to interact. Such changes give rise to a transformation of conventions and/or to an absorption of one convention by the other. This discussion offers a framework within which these phenomena can be analyzed.

In order to trace the evolution of conventions through such interactions, we formulate a non-uniform random matching model similar to the one used by Matsuyama et al. (1993), which analyzed the issues associated with international currencies. In this setup, a pair of individuals in the same society is more likely to be matched than a pair belonging to different societies. When two individuals are matched, they play one of two component games with equal probabilities. Having multiple component games to play reflects the fact that there are a variety of situations that the players have encountered, and that coordinated actions in one situation may imply misco-

---

[1]This chapter is based on Matsui and Okuno-Fujiwara (2002).

© Springer Nature Singapore Pte Ltd. 2019
A. Matsui, *Economy and Disability*, Economy and Social Inclusion,
https://doi.org/10.1007/978-981-13-7623-8_10

ordination in other situations. Introducing multiple component games enriches our analysis, as we shall see. When individuals from different societies are matched to play a component game, the action that follows a convention of one society is often not a best response to the action prescribed by the other society. A mismatch or co-ordination failure then occurs. Evolutionary pressure will change the convention in order to alleviate the coordination failure caused by a mismatch. However, a change in the convention may create another mismatch, one with the domestic convention. As this trade-off is taken into account, a new convention will evolve.

When two societies with different conventions meet, several distinct possibilities arise. First, two conventions, after some modification, may endure as two distinct conventions. Jews, Chinese, and Indians are well-known examples of worldwide merchants who depend upon their specific family network and/or methods of doing business that were taught by their parents. They keep their network/methods even if they leave their society of origin and immigrate into a new society with a different convention. Yet, in some other situations they often follow the convention of the society in which they live. These phenomena correspond to the equilibrium of our model in which two types of agent take a coordinated action in one situation and uncoordinated actions in the other.

Second, two interacting conventions may be unified. In the process of unifica-tion, often people in the small society adopt the convention of the larger society. This tendency is not surprising as many historical developments have illustrated; for example, tribal life-styles in Africa have been modified extensively by western conventions.[2]

Third, people in both societies may modify their behavior patterns to induce an eclectic convention. When Islam penetrated Menahnkabau, a part of Indonesia, there were conflicts between Islamic law and the Adat, a customary law that governed the traditional society. The conflict was rooted in the fundamental difference in the societies, i.e., patrilineal Islam and the matrilineal traditional society. After some confusion and conflicts, a new custom was established whereby wealth originated from distant ancestors (called Pusaka Tinnggi) should be inherited in accordance with the Adat, while wealth created by the immediate relatives (Pusaka Rendah) should be governed by Islamic law. This phenomenon, in which they coordinate on one convention in one situation and on another convention in other situations, is well captured in our analysis.

We examine welfare implications of these changes as well. First, if the sizes of the two societies do not differ greatly and a hybrid of the two conventions appears, as in the above example of Indonesia, welfare increases in both societies. Second, if the size of one society is sufficiently larger than the other, because the convention that evolves through interaction is the one originally adopted by the larger society, then the welfare of the smaller society increases if and only if the original convention

---

[2]Of course, what was crucial in reality was not merely the relative size of two regions, but their relative economic and military strengths.

of the larger society is superior to that of the smaller society. It should be noted, therefore, that it is mere coincidence if the smaller society's welfare increases through integration.

This chapter is organized as follows. Section 10.2 presents a two-society model and finds the equilibria of this model. Section 10.3 considers an evolutionary process to see how the people of the two societies adjust their behavior as physical integration of the two societies proceeds. Section 10.4 analyses some welfare implications. Section 10.5 extends the basic model in three directions: first, to endogenize the matching probabilities; second, to allow people to distinguish foreigners from home agents and take different strategy against them; and third, to extend the model to embrace more than two societies. Section 10.6 provides some conclusions.

10.2 The Model

Consider a world in which infinitely many anonymous and identical agents are randomly matched to play some games. Time is continuous, the horizon is infinite, and each agent is expected to match with another agent once per unit of time. In each matching, two agents play one of two component games given in Table 10.1. The game to be played is randomly assigned by Nature with an equal probability. We assume that

$$0 < \alpha, \beta < \frac{1}{2} \tag{10.1}$$

holds, i.e., that each game has two strict Nash equilibria, and that (L, L) Pareto-dominates (R, R) in game G_α, and (r, r) Pareto-dominates (l, l) in game G_β.

The entire world is divided into two societies, H (home) and F (foreign). Society H has the population fraction of $n \in (0, 1)$, while society F has $1 - n$. To express the fact that home agents meet other home agents more often than foreign agents, we consider the following non-uniform matching scheme, which is similar to the one considered in Matsuyama et al. (1993). Table 10.2 shows the probability that a row-type agent meets a column-type agent in one unit of time. In the table, $p \in [0, 1]$ is the parameter that determines the degree of integration of the two societies: $p = 0$ corresponds to an autarchy, while $p = 1$ implies complete integration.

In the main model, we assume that each agent cannot distinguish one player from another. Then a (pure) strategy of an agent is given by one of four pairs: Ll; Lr;

Table 10.1 Component games G_α and G_β

Game G_α			Game G_β		
	L	R		l	r
L	$2(1 - \alpha), 2(1 - \alpha)$	$0, 0$	l	$2\beta, 2\beta$	$0, 0$
R	$0, 0$	$2\alpha, 2\alpha$	r	$0, 0$	$2(1 - \beta), 2(1 - \beta)$

Table 10.2 Matching Scheme

	H	F
H	$1 - p(1 - n)$	$p(1 - n)$
F	pn	$1 - pn$

Rl and Rr. In this world, a (pure) strategy distribution is given by a pair of pure strategies taken by two types of agent, e.g., (Ll, Rl), where the first (resp. second) pure strategy is the one taken by the home (resp. foreign) agents. Given n, p and a strategy distribution; the payoff to a home agent who takes a certain strategy is given in the usual manner: for example, if the strategy distribution is (Ll, Rl), and if he takes Rl, then his payoff is

$$[1 - p(1 - n)]\beta + p(1 - n)(\alpha + \beta).$$

Similarly, the payoff to a foreign agent who takes Ll is

$$pn(1 - \alpha + \beta) + (1 - pn)\beta.$$

A strategy distribution, (Ss', Tt'), $(S, T \in \{L, R\}, s', t' \in \{l, r\})$, is a (pure strategy) Nash equilibrium if Ss' (resp. Tt') maximizes a home (resp. foreign) agent's payoff against (Ss', Tt').

We now characterize the set of equilibria in terms of the relative size of the societies and their degree of integration. This society has multiple equilibria. First, for all values of n, p there exists an equilibrium of the form (St, St), i.e., the agents of the both societies take the same strategy. We call such an equilibrium a unified convention equilibrium. Such a distribution is a Nash equilibrium because a strict Nash equilibrium is always played in both games.

There may exist other equilibria in which people in the two societies adopt different strategies, which we call diversified convention equilibria. We find conditions for the existence of some of these equilibria.

10.2.1   No-coordination Equilibria

The diversified equilibria can be further classified into two classes. A no-coordination equilibrium is an equilibrium (Ss', Tt') in which the home and foreign agents take different actions in both games, i.e., $S \neq T$ and $s' \neq t'$. There are potentially four no-coordination equilibria, but we analyze only (Ll, Rr) and briefly mention (Lr, Rl). Other equilibria are obtained from one of these equilibria by relabeling strategies and/or societies.

Consider (Ll, Rr). The payoff of a home agent in the equilibrium is given by

$$[1 - p(1 - n)](1 - \alpha + \beta) \tag{10.2}$$

On the other hand, if he deviates to Lr, he obtains

$$[1 - p(1 - n)](1 - \alpha) + p(1 - n)(1 - \beta). \tag{10.3}$$

Similarly, deviations to Rl and Rr will give him

$$[1 - p(1 - n)]\beta + p(1 - n)\alpha, \tag{10.4}$$

and

$$p(1 - n)(\alpha + 1 - \beta), \tag{10.5}$$

respectively. Using (10.1), we know that (10.3) is greater than (10.4). Also, when (10.5) exceeds (10.2), so does (10.3). Therefore, the incentive constraint for home agents is

$$[1 - p(1 - n)]\beta \geq p(1 - n)(1 - \beta),$$

or

$$p(1 - n) \leq \beta. \tag{10.6}$$

Similarly, the equilibrium payoff for foreign agents is

$$(1 - pn)(\alpha + 1 - \beta),$$

and the most profitable deviation is to Lr, which gives

$$pn(1 - \alpha) + (1 - pn)(1 - \beta).$$

Therefore, the incentive constraint for foreign agents is

$$pn \leq \alpha. \tag{10.7}$$

Two constraints 10.6 and 10.7 determine the equilibrium region for (Ll, Rr).

The set of pairs of n and p for which this no-coordination equilibrium exists is depicted in Fig. 10.1. What matters here is the probability of matching with agents of the other society. For a home agent, the probability of matching with a foreign agent is $p(1 - n)$, while for a foreign agent the probability of matching with a home agent is pn. The greater the probability of matching with agents of the other society, the more weight one must place on the strategy they subscribe in calculating one's best response. If the home society is relatively small ($n < \frac{\alpha}{\alpha + \beta}$), then 10.6 is more

Fig. 10.1 No-coordination
equilirium (*Ll*, *Rr*)

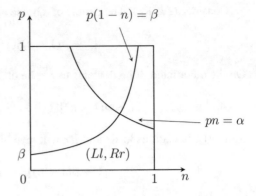

Fig. 10.2 No-coordination
equilirium (*Lr*, *Rl*)

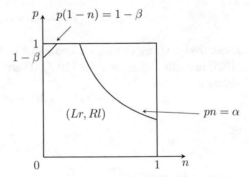

likely to be violated, and vice versa. This is because, if the home society is relatively
small, the probability of a home agent's matching with foreign agents is relatively
large, and therefore home agents are more affected by foreign agents than foreign
agents are by home agents.

The no-coordination equilibrium (*Lr*, *Rl*) has similar characteristics. The region
in which this equilibrium exists is given in Fig. 10.2. Its boundaries are given by
$p(1 - n) = 1 - \beta$ and $pn = \alpha$, which are the incentive constraints for home agents
and for foreign agents, respectively.

10.2.2 Partial Coordination

Equilibria in the other class are called partial coordination equilibria. In these equilibria, two societies coordinate on the same action in one game but not in the other.
We examine two specific partial coordination equilibria; one is (*Lr*, *Rr*), and the
other is (*Ll*, *Lr*). Other equilibria in this class can be analyzed in the same manner.

First, note that, in the strategy distribution (*Lr*, *Rr*), nobody has an incentive to
change his action in G_β since all the agents take the same action *r*. Therefore, the
only candidate for a profitable deviation for home agents is *Rr*. The payoff of a home
agent in (*Lr*, *Rr*) is

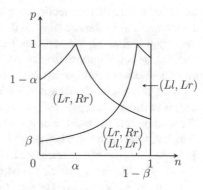

Fig. 10.3 Partial Coordination Equilibria

$$[1 - p(1 - n)](1 - \alpha + 1 - \beta) + p(1 - n)(1 - \beta).$$

On the other hand, if he takes Rr, his payoff will be

$$[1 - p(1 - n)](1 - \beta) + p(1 - n)(\alpha + 1 - \beta).$$

Thus, the incentive constraint for home agents is

$$[1 - p(1 - n)](1 - \alpha) \geq p(1 - n)\alpha,$$

or

$$p(1 - n) \leq 1 - \alpha. \tag{10.8}$$

Similarly for foreign agents the only candidate for the deviation is Lr, and the incentive constraint is given by

$$pn \leq \alpha. \tag{10.9}$$

In fact, (10.9) is the same as (10.7), and the region in which this equilibrium exists is depicted in Fig. 10.3.

The equilibrium region for (Ll, Lr) is a mirror image of that for (Lr, Rr). It is given by two incentive constraints: (10.6) and

$$pn \leq 1 - \beta. \tag{10.10}$$

This region is also depicted in Fig. 10.3.

10.3 The Evolution of Conventions Through Integration

As shown in the previous section, for any parameter values there exist multiple equilibria with qualitatively different patterns of behavior. If we look at these equilibria in a static situation, it is difficult, if not impossible, to predict which one is more likely to be found. This section employs the best response dynamics as a selection device to identify the equilibrium that emerges when the world starts with a certain strategy distribution. We suppose that in the beginning there is no physical interaction, i.e., $p = 0$. Suppose further that the initial distribution is (Ll, Rr), which is the no-coordination equilibrium examined in the previous section.

With this initial distribution, we look at an evolutionary process when p, the degree of integration, gradually increases with n being fixed throughout the process. We assume that the change in p is sufficiently slow that any adjustment in strategy distribution is completed before p changes further. We analyze the case of $n < \frac{\alpha}{\alpha+\beta}$. The opposite case can be analyzed in the same manner and will be illustrated as well. If $n < \frac{\alpha}{\alpha+\beta}$, the first constraint to be violated through the process of increasing p is $p(1 - n) \leq \beta$. Once p goes beyond this constraint, the incentive constraint for home agents is violated, and they start taking Lr. This process continues until all the home agents take Lr, and the partial coordination equilibrium (Lr, Rr) emerges. To see a further change, the case $n < \frac{\alpha}{\alpha+\beta}$ ought to be divided into two subcases: (i) $n < \alpha$ and (ii) $\alpha < n < \frac{\alpha}{\alpha+\beta}$, where we ignore the boundary.

In (i), when the home society is very small compared with society F, then, as p increases further, it is again the home agents' incentive constraint that is violated first. This happens when p increases beyond constraint (10.8). In this case, each home agent has an incentive to switch his strategy to Rr. The unified equilibrium in which everyone uses the convention of the foreign society emerges. In other words, the convention of the home society is absorbed into that of the foreign society.

In subcase (ii), after reaching the partial coordination equilibrium (Lr, Rr), it is now the F agents that will switch their strategies. As we saw in the static analysis, they change their strategy to Lr when p goes beyond α/n. Unification also occurs. But in this case the two conventions are mixed: in G_α, agents follow the convention originally established in the home society, while in G_β, agents follow the convention of the foreign society. These cases, and the cases when $n > \frac{\alpha}{\alpha+\beta}$, are shown in Fig. 10.4.

Note that the process of adjustment is irreversible in the sense that, once coordination/assimilation occurs, the society cannot retrieve its old convention even if p decreases. The society loses its custom forever.

Fig. 10.4 Integration and equilibria

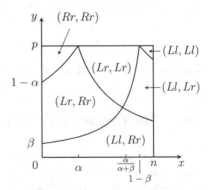

10.4 Welfare

This section examines some welfare implications, especially those of integration. First of all, in unified equilibria the equilibrium payoffs to an agent, common to both types, are as follows:

$$
\begin{array}{ll}
\text{Equilibrium Payoff} & \\
(Ll, Ll) & 1 - \alpha + \beta \\
(Lr, Lr) & 2 - \alpha - \beta \\
(Rl, Rl) & \alpha + \beta \\
(Rr, Rr) & \alpha + 1 - \beta
\end{array}
\tag{10.11}
$$

Under our assumptions on α and β (that both are less than a half), Lr is the optimal strategy to coordinate on. Consider the evolutionary process examined in the previous section with the initial condition of (Ll, Rr); and suppose that p goes to 1 in the limit. Then, if n is between α and $1 - \beta$, i.e. if the sizes of the two societies are not too different, the integration will bring an improvement to both societies. Note that this result will not be obtained if the world started with two similar conventions. Indeed, if the two societies both started from the same convention, then there will be no further change. The world with diversified conventions attains a high level of welfare in the end.

If one society, say, society H, is sufficiently small compared with the other, or more specifically if $n < \alpha$, then the integration leads to (Rr, Rr), the convention originally established in society F. There is no guarantee that the welfare in this situation will be higher than the welfare in the original one–indeed, it is determined only by the relative size of α and β. If $\alpha < \beta$, then the H agents will be worse off as the result of integration. This indicates that a statement such as "One convention absorbs the other because the former is better than the latter" is too simplistic a view.

10.5 Extensions

This section considers two modifications of the model developed in the previous sections. The first introduces the possibility of differentiating between H-type and F-type agents when taking strategies. The second is a model with more than two societies.

10.5.1 Endogenous Matching Probability

In the real world, matching is not completely exogenous. People often choose which group of people they want to meet with. Among various specifications, this subsection considers the following. Suppose that each agent can choose the propensity to match with agents in the other society: $p_h \in [0, 1]$ for an H agent, and $p_f \in [0, 1]$ for an F agent. We allow agents in the same society to choose different propensities. Let $\bar{p}_h$ (resp. $\bar{p}_f$) be the average propensity of H agents (resp. F agents). In order for the matching technology to be consistent, we assume that the actual matching probability with the other group is

$$\frac{p_h + \bar{p}_f}{2}(1 - n)$$

for an H agent who chooses p_h, and

$$\frac{p_f + \bar{p}_h}{2}n$$

for an F agent who chooses p_f. The strategy of an agent is written as (Ss', p'). Also, we assume that they can choose the propensity with no additional cost.

Suppose now that the initial state is an autarky state, i.e., $\bar{p}_h = \bar{p}_f = 0$, with (Ll, Rr). It can be easily verified, using the payoff structure and the linearity of the expected payoff in p_h, that the only possibilities for a best response of an H agent are $(Ll, 0)$, $(Rr, 1)$ and $(Lr, 1)$. Several results are immediate. First, an H agent obtains the payoff of $1 - \alpha + \beta$ in the initial state. Second, if he chooses $(Rr, 1)$, the probability of his matching with F agents is $\frac{1}{2}(1 - n)$ since $\bar{p}_f = 0$ holds. Therefore, his expected payoff will be

$$\frac{1}{2}(1 - n)(\alpha + 1 - \beta). \tag{10.12}$$

Third, if he chooses $(Lr, 1)$, then his expected payoff becomes

$$\left(1 - \frac{1}{2}(1 - n)\right)(1 - \alpha) + \frac{1}{2}(1 - n)(1 - \beta). \tag{10.13}$$

Subtracting (10.12) from (10.13), we obtain

$$\left(1 - \frac{1}{2}(1 - n)\right)(1 - \alpha) - \frac{1}{2}(1 - n)\alpha = (1 - \alpha) - \frac{1}{2}(1 - n).$$

Then $\alpha < 1/2$ implies that this expression is always positive. Therefore, $(Rr, 1)$ cannot be a best response.

Next, from the initial state payoff and (10.13), $(Lr, 1)$ is the best response to the initial state if and only if

$$\frac{1}{2}(1 - n)(\alpha - \beta) \geq \beta. \tag{10.14}$$

A symmetric analysis gives us the result for F agents' behavior: $(Lr, 1)$ is the best response to the initial state for an F agent if and only if

$$\frac{1}{2}n(\beta - \alpha) \geq \alpha. \tag{10.15}$$

If neither (10.14) nor (10.15) holds, then the initial state $((Ll, 0), (Rr, 0))$ continues forever. In particular, if $1/2 < \alpha/\beta < 2$, no evolutionary change will occur regardless of the relative size of the societies.

On the other hand, if (10.14) holds, then H agents start taking $(Lr, 1)$. Once this process starts, $(Lr, 1)$ becomes more appealing, until it reaches the state where every H agent takes $(Lr, 1)$. After the society reaches the state $((Lr, 1), (Rr, 0))$, the incentive for F agents to change their behavior may arise. The only candidate for a best response, other than the current strategy, is $(Lr, 1)$. First, in the present state, the expected payoff of an F agent is

$$\frac{1}{2}n(1 - \beta) + \left(1 - \frac{1}{2}n\right)(1 - \beta + \alpha). \tag{10.16}$$

Second, the expected payoff from taking $(Lr, 1)$ is

$$n(1 - \alpha + 1 - \beta) + (1 - n)(1 - \beta). \tag{10.17}$$

Subtracting (10.16) from (10.17), we obtain

$$n - \alpha.$$

Therefore, if $n < \alpha$, then $((Lr, 1), (Rr, 0))$ becomes an absorbing state. On the other hand, if $n > \alpha$, then $(Lr, 1)$ also becomes a best response for F agents. The more F agents take this strategy, the greater incentive they have to follow it. The strategy distribution reaches $((Lr, 1), (Lr, 1))$. This distribution becomes the absorbing state of the best response dynamics.

Unlike the case of exogenous matching technology, people will always move towards a superior convention if they actually change their behavior. However, it is more likely in this case that people will avoid interacting with foreigners, which will lead to no improvement in conventions. This will occur even when the hybrid convention is far better than the convention established in either society, e.g., when α and β are both close to zero.

10.5.2 Government Intervention

As the discussion in the previous sections illustrates, interaction with another society may harm the home society's welfare, not only because the short-run effect creates coordination failures associated with the matching of home with foreign agents who play differently in the component games, but also because the home agents' strategies may be unfavorably affected by the foreign agents' strategies through evolution. The extreme case of such phenomena occurs if the home society's strategy is initially optimal (i.e., Lr), while the foreign society's strategy is suboptimal, say, Rr. To make the result stark, suppose further that the relative size of the home society is almost negligible ($n = \varepsilon$), and that p goes from 0 to 1. In such a case, in the short run home agents will suffer from miscoordination with foreign agents in G_α, while in the long run home agents' strategies will converge to Rr.

Faced with such a possibility, the home government may have an incentive to intervene. For example, it may be able to effectively close the society by forbidding any interaction with foreign agents. By doing so, it can avoid coordination failures in the short run and preserve the optimal convention in the long run. Alternatively, it may be able to impose a tax/penalty for domestic agents who take some specific action, if taking the action is verifiable. For example, if it can impose a tax $\gamma > 0$ for any player taking action R in G_α, the best response will be altered from Rr to Lr under a certain parameter set. By appropriately choosing γ, the government can preserve the home society's convention even after interaction begins.[3]

All of the above results, however, critically depend upon the amount of information possessed by the government. In particular, the government is likely to have as little information as individual agents have, and hence is likely to be myopic. A myopic government is likely to judge that interaction with foreign societies is detrimental to the home society because it is the loss resulting from miscoordination with foreign agents in the short run that tends to dictate its judgement, rather than the potential merit of a change in domestic convention in the long run.

10.5.3 Discrimination

If it is not too costly to take different actions against different opponents, agents may consider such a possibility. In addition to the original two-society model, assume that an agent can discern home agents from foreign agents and take different strategies against them if he pays $d > 0$ per game. The cost is either associated with discernibility of the two types of agent or is a cost of holding two strategies/options available. Denote by (Ss', Tt') the strategy of a single agent who takes Ss' against a home agent and Tt' against a foreign agent. We also identify (Ss', Tt') with the strategy distribution in which every individual in one society takes this strategy. We consider an evolutionary process in which p changes very slowly relative to the speed of adjustment of the behavioral pattern of the society. When the society is at $((Ss', Tt'), (S^*s^{*'}, T^*t^{*'}))$, it is in the best interest of a home agent (resp. foreign

---

[3]This policy may create some distortion, even if the tax revenues are returned to the domestic agents in the form of per capita subsidy. Again, the net effect may very well be negative.

agent) to take $(Ss', S^*s^{*'})$ (resp. $(Tt', T^*t^{*'})$) on the condition that he pays d. We assume $n < 1/2$ throughout the rest of the analysis.

We start with the initial condition $((Ll, Ll), (Rr, Rr))$, which corresponds to the initial condition focused on in the previous sections. By taking a discriminatory strategy, an H agent gains

$$p(1 - n)(\alpha + 1 - \beta). \tag{10.18}$$

at the expense of d. Note that (10.18) is equal to zero at $p = 0$ and increasing in p. Therefore, if $d > 0$ is not too large, there exists a minimum $p > 0$, denoted by p_d, at which (10.18) is equal to d. Assume that this is the case, and that every home agent takes a discriminatory strategy (Ll, Rr). Once we obtain the strategy distribution $((Ll, Rr), (Rr, Rr))$, no further change in strategy occurs even if p changes, since every pair of agents plays a strict Nash equilibrium in every game.

In the above case, discrimination may not be a good strategy in the long run from the viewpoint of the society as a whole, as it prevents further adaptation of convention. Suppose $n \in (\alpha, 1 - \beta)$. Then either home agents or foreign agents will start to discriminate against agents of the other society in order to avoid a loss from mismatch; but the convention will not change any further, and potential gains from achieving integration and adopting Lr will be lost. Public intervention may be called for in order to disallow discrimination if the government has sufficient knowledge to be able to foresee correctly the future course of adaptation.

10.5.4 A K-Society Model

We can extend our analysis to a world with more than two societies. In such a model the order of integration may matter. To see this, suppose that there are three societies, 1, 2 and 3, with equal sizes of $1/3$. Let α be greater than $1/3$. Assume further that there is no interaction between any two societies in the beginning, and that society 1 starts with Ll, while societies 2 and 3 start with Rr. If societies 1 and 2 first experience integration, then they will end up with Lr since they are of equal size. If, after this integration, the society made of 1 and 2 (call it society 1–2) and society 3 are integrated, the world will end up with Lr. On the other hand, if societies 2 and 3 are integrated first and then this amalgamated society is integrated with society 1, the world will end up with Rr. This is because when society 1 and society 2–3 meet, their sizes are $1/3$ and $2/3$, respectively.

To get the starkest result, suppose that there are K societies. Then, even if Ll is adopted by a single society, 1, with a size of $n_1 = 1/K$, and if the rest of societies, $2, \ldots, K$, adopt Rr, the world may end up in Ll. To see this, let the size of society $k = 2, \ldots, K$ be given by

$$n_k = \frac{K^{1/(K-1)} - 1}{K} K^{(k-2)/(K-1)}.$$

Note that we have $n_2 < n_3 < \cdots < n_K$ and $\sum_{k=1}^{K} n_k = 1$. Now, suppose that society 1 and society 2 are integrated first, that the integrated society absorbs society 3, and so on in such an order that society $k = 2, 3, \ldots, K$ is integrated by the compound society consisting of societies $1, 2, \ldots, k - 1$. Given that we constructed the example so as to satisfy $\sum_{\ell=1}^{k} n_\ell : n_{k+1} = K : K^{1/(K-1)} - 1$, the ratio converges to one-to-zero as K goes to infinity. Therefore, for a sufficiently large K, the society will end up with Ll. On the other hand, if societies $2, \ldots, K$ are integrated first, then the final outcome will be Rr for a sufficiently large K. In this example, the order of integration determines the final outcome of the world.

10.6   Remarks

We have analyzed how societies with different conventions would interact in an evolutionary environment. Our model is somewhat restricted. For example, members of two societies interact with each other directly, not through communication media, and the component games are simple coordination games with only two possible actions. However, within such a simple framework we have shown some interesting results. There are various patterns of evolutionary paths: sometimes one convention absorbs the other, sometimes two original conventions are preserved without any change, while in other cases a new convention may evolve as a result of interaction. Whichever pattern emerges depends on, among other things, the relative sizes of the societies and the extent of their interaction. It may happen that the convention of the smaller society is superior to that of the larger, and if that is the case the smaller society will be worse off as a result of assimilation into the larger society.

   If the matching technology is endogenously determined, people will always move towards a superior convention if they actually change their behavior. However, it is more likely in this case that people avoid interacting with foreigners, which leads to no improvement in conventions. This happens even when the hybrid convention is far better than the convention established in either society.

   An additional result of our analysis, which is implicit in the model, may be worth emphasizing. As many analyses of coordination games have illustrated, a society is often trapped in an inefficient equilibrium, and escaping from such a trap is difficult. If we wait for the private sector to take the initiative, it may take a long time for something to happen even if the system is shaken constantly (Foster and Young 1990; Kandori et al. 1993; Young 1993), or even if a secret handshake (Robson 1990) or communication (Chap. 9, or Matsui 1991) is allowed.

   There are several ways to accelerate a departure from such a trap, even if the matching structure is not local (Ellison 1993). First, as illustrated by many policy-oriented analyses (see e.g., Okuno-Fujiwara 1988), the government may use subsidies and taxes to alter the incentives of the private sector. Such a policy, however, may require a lot of information and vast resources to implement. The second way is to create a euphoric expectation about the future and thereby let people coordinate on an efficient outcome (Matsuyama 1991; Krugman 1991).[4] The success of this type of policy depends on how private sectors react to the government propaganda.

In this analysis, we have identified yet another possibility for coordination. Like artificially creating a hybrid by crossing two different genes, an interaction of societies affects the conventions of each society and sometimes creates a hybrid of conventions, forcing the society to escape from the trap. It should be emphasized again, however, that this happens only when the two societies are of similar size. If one society is sufficiently larger than the other, integration will lead to a complete absorption of the convention of the smaller society by that of the larger one. Welfare may move in either direction if that is the case.

---

[4]Matsui and Matsuyama (1995) show that in a coordination game, a risk dominant outcome is chosen by such a process.

Chapter 11
When Trade Requires Coordination

11.1 Introduction

This chapter considers an economy in which coordination on the way interaction occurs is required in order to make transactions. Coordination on the way two parties communicate is a typical example. In terms of language, coordination between English and Japanese, or between oral and sign languages, often induces such an issue.

Interaction requires coordination, and the interactions that arise in economic activities are certainly no exception.[1] International trade, for example, requires coordination on a variety of conventions including the language used in negotiations. Corporate mergers often require that the merging companies coordinate on a single corporate culture. Network externalities force agents to coordinate on one or, at most, a small number of operating systems. Rural to urban migration results when people must coordinate on a geographical location for many nonagricultural sector jobs. Less obvious, but nonetheless important, examples can be found in behavioral assimilation of minority groups into majority groups. Immigrants often assimilate into the predominant culture of their new country, or phrased differently, they coordinate their customs with those of their new compatriots. Members of racial or ethnic groups that have historically interacted largely with only members from their own community find that a similar predicament applies to them. This chapter studies the effects of integration of two communities in the presence of such need for coordination. It provides us with a result, albeit in a limited situation, against the received wisdom that trade enhancing integration always increases the welfare of the two communities. In the sense that we do not present general and comprehensive results, the purpose of this chapter is suggestive rather than definitive in nature.

If two communities are socially and/or economically integrated, members of a minority group may have the incentive to adopt behaviors of a majority group even if they inherently prefer their own behavior patterns. This is because adopting these

---

[1] This chapter is based on Katz and Matsui (2004).

© Springer Nature Singapore Pte Ltd. 2019
A. Matsui, *Economy and Disability*, Economy and Social Inclusion,
https://doi.org/10.1007/978-981-13-7623-8_11

alternative behaviors allows them to interact with a larger group of people. This force for coordination is referred to as strategic complementarity, and its effects have been studied extensively in development economics, international economics, and game theory. However, welfare analysis is somewhat limited. Only gains from trade and coordination are often mentioned, and focuses are on how coordination on a desirable outcome can be achieved through various measures.

The observation which we incorporate into the present model is the fact that different agents may have different inherent preferences over a set of standards on which they coordinate. For instance, speaking a foreign language requires an additional effort compared to using one's native language. The expenses of moving one's home from one location to another consist not only of direct moving expenses but also of various adjustment costs. Similarly, changing one's behavior patterns to those that are less natural or instinctive can be uncomfortable or even traumatic, as is eloquently phrased by Asimov (1995):

> I am the son of immigrant parents. [. . .] I had to learn American culture on my own, and it
> is difficult to explain to someone who has not gone through it what this means. . . . In short,
> I was a cultural orphan (like many others) and "speak the culture with an accent."

Incorporating such heterogeneity in inherent preferences over standards of behavior, we study the relationship between the need for coordination and the standards of behavior that people choose. In particular, we explicitly study a dynamic process and some welfare implications. In so doing, we find that when a small community is integrated with a large community, the small community ends up worse off than before provided that standard gains from trade based on comparative advantage are sufficiently small.

Roughly speaking, we construct a random matching model of trade with the following characteristics. First, gains from trade are higher if two agents use the same standard of behavior. Each agent chooses such a standard from an exogenously given set of standards before she is randomly matched with a trading partner. If a matched pair of agents have chosen the same standard, and if a beneficial trade exists, then they will trade their goods. If they have not chosen the same standard, then trade may be possible, but agents incur a significant cost due to the lack of coordination. We normalize this payoff to zero in the model.

Second, all members of society belong to one of two distinct communities with different sizes, with members of one community preferring one standard while members of the other community prefer another. The costs of adopting a less preferred standard may differ among agents within a community. We focus our analysis on the situations in which one country is smaller than the other in terms of the production of the two goods, so that the former is unambiguously smaller than the latter. This assumption, albeit not universal, is not unrealistic. One may think of situations such as trades between the United States and Kuwait where the United States produces far more crude oil than Kuwait (USA, 39,304; Kuwait, 11,345; thousand metric tons, average per month in 2017: United Nations 2019), but the former imports oil from the latter.

As a reference point, we begin with a situation in which agents trade only with members of their own community. This corresponds to situations in which there is a barrier between the two communities, in the sense that a member of one community simply does not come into contact with members of the other community. We assume that each community is in the equilibrium in which all members choose the standard preferred within that community. We then study a dynamic process and its long-run result after this barrier is lifted. It is shown that there is a unique equilibrium that is accessible from the initial autarkic situation under any monotone dynamic, and that in the process of moving toward the new equilibrium, members of the smaller community assimilate into the larger community.

We make welfare analysis, comparing autarky and free trade, which is made possible by the uniqueness of the accessible outcome after the barrier is lifted. In the main part of welfare analysis, we assume that the matching technology exhibits constant returns to scale. In this case, there are essentially two effects of integration. If there is no inherently preferred standard, then integration increases the welfare of the both communities due to standard gains from trade based on uneven distributions of endowments. This is the positive effect of integration. On the other hand, due to distinct inherent preferences over a set of standards, those who assimilate into the larger community incur costs, which induces a negative effect. The total effect of integration is determined by the relative size of these two effects. If gains from trade are minimal, then the loss caused by adopting the less preferred standard becomes a dominant factor, and total welfare decreases when the barrier is lifted.

We also analyze the case in which the matching technology exhibits increasing returns to scale. In this case, integration increases the matching probability for each agent. This scale economy induces a positive effect on welfare when the communities are integrated. Therefore, nonexistence of gains from trade based on comparative advantage does not provide a sufficient condition for a decrease in welfare after integration. However, there are nontrivial situations in which integration leads to a welfare loss.

Under the assumption of increasing-returns-to-scale matching technology, we also look briefly at a situation in which agents may choose from a set of more than two standards. We offer an example in which, beginning from the autarkic situation in which all agents use their respective most preferred standards, all agents in both communities change their standard to a third standard when the barrier between the communities is lifted. This third standard is not the most preferred standard of either community but is most "easily accessible" by both. We identify some conditions under which all members in both communities are left worse off.

The present analysis is built on the framework used in the previous chapter, which considers a situation in which players from two regions are matched to play some coordination games. It analyzes conditions under which complete and partial assimilation occur as a result of integration of the two regions. It is also shown that eclectic (hybrid) behavior may arise as a result of integration.

The present analysis is also related to Lazear (1999), which examines which minority group tends to learn English faster than others after migration to the United States. It is shown that the larger the size of the group is, the slower people in the group learn it. Members of a smaller group may find an more urgent need to coordinate with the majority of the society than those of a larger group. Strategic complementarity is the main source to account for this phenomenon. The present analysis goes beyond these works to claim that as a result of such assimilation, the minority group may be worse off.

Another paper related to the present analysis is Akerlof and Kranton (2000). Different inherent preferences over a set of standards can be interpreted as "identities" in their terminology. Several equilibria are compared when two groups of people interact with each other. The main difference between their paper and the present analysis lies in the fact that we identify a single equilibrium that is accessible from the autarky equilibrium, and thereby make it possible to compare welfare before and after a trade barrier is lifted.

The logic we use in the analysis to show that one community may be worse off when the barrier is lifted differs from that found in Hart (1975) and in the literature on customs unions, both of which also examine cases in which a welfare loss can result in the lifting of a trade barrier. First, Hart examines a case involving incomplete markets and shows that the addition of an asset which allows trade between some of the markets, but not all of them, can lead to a decrease in welfare. Second, in the literature on customs unions, it is shown that when a country has two trading partners, one with high costs and the other with low costs, and this country forms a customs union with the higher cost partner, while imposing a high tariff on the low-cost partner, then a decrease in welfare is also possible. Both examples essentially claim that a "partial shift" toward the first best does not necessarily improve welfare. Instead of using this logic, our welfare result relies on the negative externality induced by a switch from a favorable standard to an unfavorable one.

The remainder of this chapter is organized as follows. The formal model is laid out in Sect. 11.2. Section 11.3 characterizes equilibria, which is followed by the analysis of dynamics in Sect. 11.4. An analysis of the welfare implications that result from our accounting for the cost of coordination makes up Sect. 11.5. Section 11.6 assumes that the matching technology exhibits increasing returns to scale, and continues our welfare analysis. Although the welfare implications are not as clear-cut as in the case of constant returns to scale, two examples are offered to show a decrease in welfare after integration. Section 11.7 concludes the analysis.

11.2   The Model

We consider an exchange economy consisting of two types of agents, type 1 and type 2. The sizes of the type 1 group and type 2 groups are equal. There are two indivisible commodities, 1 and 2. A type 1 agent is endowed with two units of good 1, while a type 2 agent is endowed with two units of good 2. Each type 1 agent is indexed by a number in $(0, 1)$, as is each type 2 agent. Agents get zero utility from consuming only one good, while they get positive utility from consuming a unit each of 1 and 2. In this world, trade takes place in the form of a one-for-one swap of goods. Agents cannot trade their goods unless they coordinate on one of two behavioral standards, L and R.

Each agent belongs to one of two communities, A and B. Agents in A prefer coordination on standard L, while agents in B prefer coordination on R. However, coordination on some standard is preferred to miscoordination by all agents. If an agent i of type 1 (resp. 2), from community A, uses standard L and trades with an agent of the opposite type using L, the agent receives a utility level that we normalize to 1, while if he uses standard R and trades with another agent using R, he receives utility $\mu_R(i) \in [0, 1]$ (resp. $\nu_R(i) \in [0, 1]$). Similarly, if an agent j of type 1 (resp. 2) from community B uses standard R and trades with an agent of the opposite type using R, the agent receives utility 1, while if he uses standard L and trades with another agent using L, he receives utility $\mu_L(j) \in [0, 1]$ (resp. $\nu_L(j) \in [0, 1]$).

We denote the fraction of the type 1 agents in community A by m and the fraction of the type 2 agents in community A by n. Thus, the fraction of the total population of community A is $(m + n)/2$. We arrange type 1 agents uniformly on the line $[0, 1]$ in the following manner: for $i \leq m$, $0 \leq \mu_R(i) \leq 1$ and $\mu_L(i) = 1$; for $i > m$, $0 \leq \mu_L(i) \leq 1$ and $\mu_R(i) = 1$. In addition, $\mu_R(i)$ is (weakly) increasing, and $\mu_L(i)$ is (weakly) decreasing. This places players in community A on the first portion of the line and those in B after them. Also, note that this is equivalent to placing the players who have the most trouble changing their standard of behavior at the outer ends of the line, while placing those who would find it least difficult to make this change nearest the members of the community to which they do not belong. In the same manner, we arrange type 2 players on the line $[0, 1]$ as follows: for $j \leq n$, $0 \leq \nu_R(j) \leq 1$ and $\nu_R(j) = 1$; for $j > n$, $0 \leq \nu_L(j) \leq 1$ and $\nu_L(j) = 1$; $\nu_R(j)$ is increasing, and $\nu_L(j)$ is decreasing.

There is no centralized market where agents can meet to exchange commodities. Rather, agents are randomly matched into pairs. We examine two cases. We will refer to the first of these cases as the "autarky" case, which will serve as a benchmark for the second case (the "unification" case). In the autarky case, agents are matched only with agents from their own community. Matching is uniform within each community, and the matching technology exhibits constant returns to scale. In other words, a type 1 (resp. type 2) agent in community A meets a type 2 (resp. type 1) agent with probability $n/(m + n)$ (resp. $m/(m + n)$). Similarly, a type 1 (resp. type 2) agent in community B meets a type 2 (resp. type 1) agent with probability $(1 - n)/(2 - m - n)$ (resp. $(1 - m)/(2 - m - n)$).

The second case is the unification case. Here, agents may meet trading partners from either community. Matching is uniform. In this case, an agent will meet a trading partner (i.e., an agent of the opposite type) with probability $1/2$. Recall, however, that this is not equivalent to saying that an agent will trade with probability $1/2$. Agents in a pair must use the same standard of behavior.

The fractions of agents who are matched with agents of the opposite type are $2mn/(m + n)^2$ in the autarky case, and $1/2$ in the unification case, respectively. This implies that if $m \neq n$, i.e., if each community has a comparative advantage in one of the goods, then the fraction of matches between two distinct types of agents is greater in the unification case than in the autarky case. Gains from trade would arise if it were not for the need for coordination.

11.3 Equilibria

11.3.1 Autarky

In the autarky case, since no agent can meet an agent from the other community, we can analyze each community separately. We consider community A. The analysis for community B is symmetric. First, any pure strategy equilibrium can be characterized by two numbers, $x \in [0, m]$ and $y \in [0, n]$, where agent i of type 1 (respectively type 2) takes standard L if and only if $i < x$ (respectively $i < y$). Indeed, if $i < j$, then $\mu_R(i) \leq \mu_R(j)$ and $\nu_R(i) \leq \nu_R(j)$ and therefore, if player i takes R in an equilibrium, player j of the same type weakly prefers R to L as well.

On average, the agent i of type 1 obtains $y/(m+n)$ if he takes L and $\frac{n-y}{m+n}\mu_R(x)$ if he takes R. Therefore, his incentive conditions are given by

$$y \geq (n - y)\mu_R(i) \text{ if } i < x, \text{ and}$$
$$y \leq (n - y)\mu_R(i) \text{ if } i > x.$$

These inequalities give the incentive curve for type 1 agents, i.e., the curve on which no agent of type 1 has an incentive to deviate:

$$y = (n - y)\mu_R(x). \tag{11.1}$$

Similarly, for an agent i of type 2, we have

$$x \geq (m - x)\nu_R(i) \text{ if } i < y, \text{ and}$$
$$x \leq (m - x)\nu_R(i) \text{ if } i > y.$$

These inequalities give the incentive curve for type 2 agents:

$$x = (m - x)\nu_R(y). \tag{11.2}$$

The intersections of (11.1) and (11.2) determine the equilibria of this community. Since $\mu_R(x)$ and $\nu_R(y)$ are functions of x and y, respectively, we can draw equilibrium conditions on a (x, y)-plane.

There are always two equilibria, $(0, 0)$ and (m, n) since there is no gain from trade by taking a standard that is taken by nobody. Two examples are given in Figs. 11.1 and 11.2. The first illustrates the case in which μ and ν are distributed uniformly, i.e.,

$$\mu_R(i) = \frac{i}{m}, \tag{11.3}$$

and

$$\nu_R(i) = \frac{i}{n}. \tag{11.4}$$

Fig. 11.1 Phase diagram: autarky with uniform distribution

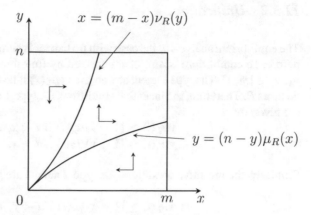

Fig. 11.2 Phase diagram: autarky with two mass points

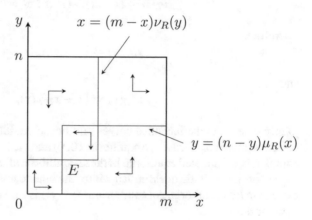

On the other hand, the second illustrates the case where the distributions are made up of two mass points at 0 and 4/5 with a quarter of each type having 0, i.e.,

$$\mu_R(i) = \begin{cases} 0 & \text{if } i \le \frac{1}{4}m, \\ \frac{4}{5} & \text{if } i > \frac{1}{4}m, \end{cases} \tag{11.5}$$

and similar for $\nu_R(\cdot)$. The arrows in the figures show their incentives.

In the second example, there are equilibria other than the two equilibria, $(0, 0)$ and (m, n). Outcome $E = (\frac{1}{4}m, \frac{1}{4}n)$ is one of them. In this equilibrium, those who obtain no utility from coordinating on R stick to standard L, while those who obtain a positive utility take standard R. The community is divided into two in this equilibrium.

11.3.2 Unification

The equilibrium analysis of the case with no barrier is similar to that of the autarky. As before, an equilibrium is now characterized by four thresholds, $x_A, x_B \in [0, m]$ and $y_A, y_B \in [m, 1]$. The type 1 agent at x obtains $y\mu_L(x)$ if he takes L, and $(1 - y)\mu_R(x)$ if he takes R. Therefore, the incentive conditions for type 1 agents in both communities are given by

$$y\mu_L(i) \geq (1 - y)\mu_R(i) \text{ if } i < x, \text{ and}$$
$$y\mu_L(i) \leq (1 - y)\mu_R(i) \text{ if } i > x.$$

Similarly, the incentive conditions for type 2 agents are given by

$$x\nu_L(i) \geq (1 - x)\nu_R(i) \text{ if } i < y, \text{ and}$$
$$x\nu_L(i) \leq (1 - x)\nu_R(i) \text{ if } i > y.$$

Therefore,

$$y\mu_L(x) = (1 - y)\mu_R(x), \tag{11.6}$$

and

$$x\nu_L(y) = (1 - x)\nu_R(y), \tag{11.7}$$

jointly determine the incentive curves and hence equilibrium. For every set of parameters, there are at least two equilibria, $(0, 0)$ and $(1, 1)$. These equilibria are called completely assimilated equilibria. In these equilibria, all members of one community have coordinated with, or assimilated into, the other community. In general, there are more than two equilibria. An equilibrium (x^*, y^*) is said to be a *partially assimilated equilibrium* if

$$(x^*, y^*) \in [0, m] \times [0, n] \cup [m, 1] \times [n, 1] \setminus \{(0, 0), (m, n), (1, 1)\}.$$

Here, some members of one community have coordinated with the members of the other community, but not all members have done so. One community is divided due to integration with the other community.

Figures 11.3 and 11.4 illustrate two examples (ignore arrows for the moment). Figure 11.3 corresponds to the case of uniform distribution given by (11.3) and (11.4) for community A, and symmetrically,

$$\mu_L(i) = \frac{1 - i}{1 - m}, \quad i > m, \tag{11.8}$$

and

$$\nu_L(i) = \frac{1 - i}{1 - n}, \quad i > n, \tag{11.9}$$

for community B. In the figure, we let $m = 0.4$ and $n = 0.25$ for the sake of illustration. Recall that $\mu_L(i) = 1$ holds for $i \leq m$, and so on.

Figure 11.4 corresponds to the case where there are two mass points in each community in terms of the distributions of μ and ν. There is a seemingly stable partially assimilated equilibrium, but in order to see the stability of equilibria and which equilibrium may arise as the result of lifting the barrier, we must turn to dynamical analysis.

Fig. 11.3 Phase diagram: uniform distribution

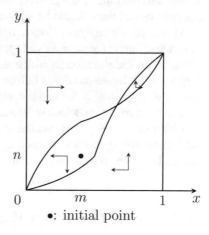

•: initial point

Fig. 11.4 Phase diagram: two mass points

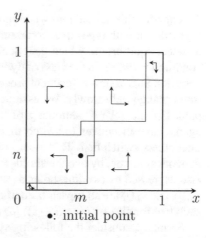

•: initial point

11.4 Dynamics

We study dynamics after the barrier is lifted. To this aim, we assume that time is continuous, and the horizon is infinite. Each agent of type 1 (resp. 2) can produce two units of good 1 (resp. 2) at a time right after she consumes a unit each of the two goods. In general, we do not know which equilibrium emerges as the result of integration if there are multiple equilibria. Indeed, it is the nature of conventions/standards that the outcome is history dependent. If the two communities coordinate on the same standard at the time of integration, then this standard continues to be used after the integration, and there would be no issue of assimilation.

Therefore, we analyze below a more interesting case than this, i.e., the situation in which different communities adopt different standards. In particular, we focus our attention on the situation in which agents in each community coordinate on their respective preferred standards before the barrier is removed. It corresponds to the case in which (m, n) is the initial condition.

The first subsection, which is the main part of the present section, studies situations in which community A is smaller than community B in both types. The second subsection briefly looks at situations in which community B contains more type 1 agents, but fewer type 2 agents, than community A.

11.4.1 Small Community Versus Large Community

Throughout this subsection, we assume $m < 1/2$ and $n < 1/2$, i.e., community A is smaller in both types than community B. We use a class of dynamical processes to select equilibrium, which generically makes our results conclusive if the initial condition is (m, n). In the class of dynamical processes, agents gradually adjust their behavior. Such a slow adjustment process is appropriate in our problem, since cultural traits change only slowly. We assume that time is continuous, as is the dynamical path. For the sake of simplicity of the analysis, we assume further that if many agents have incentives to switch their actions, those who have greater incentives than others switch first. This assumption enables us to characterize the state of the dynamical system by two thresholds x and y on condition that the initial condition is also expressed by two thresholds as we have assumed. We assume that $\mu_R(\cdot)$, $\nu_R(\cdot)$, $\mu_L(\cdot)$, and $\nu_L(\cdot)$ are all Lipschitz continuous, i.e., for all i, there exist $\varepsilon > 0$ and $K > 0$ such that for all $j \in (i - \varepsilon, i + \varepsilon)$, $|\mu_R(i) - \mu_R(j)| < K|i - j|$ holds, and so forth.

Formally, consider the following system of differential equations:

$$\begin{cases} \dot{X} = F(X, Y) \equiv f(X, \mu_L(X)Y - \mu_R(X)(1 - Y)), \\ \dot{Y} = G(X, Y) \equiv g(Y, \nu_L(Y)X - \nu_R(Y)(1 - X)), \end{cases} \tag{11.10}$$

where f and g are Lipschitz continuous, continuously differentiable, (weakly) increasing in the second argument, and satisfy the following:

$$
\begin{aligned}
f(X,0) &= g(Y,0) = 0, \\
f(X,Z) &> 0 && \text{if } Z > 0, X \neq 1, \\
f(1,Z) &= 0 && \text{if } Z > 0, \\
f(X,Z) &< 0 && \text{if } Z < 0, X \neq 0, \\
f(0,Z) &= 0 && \text{if } Z < 0, \\
g(Y,Z) &> 0 && \text{if } Z > 0, Y \neq 1, \\
g(1,Z) &= 0 && \text{if } Z > 0, \\
g(Y,Z) &< 0 && \text{if } Z < 0, Y \neq 0, \\
g(0,Z) &= 0 && \text{if } Z < 0.
\end{aligned}
$$

In (11.10), $\mu_L(X)Y - \mu_R(X)(1-Y)$ (resp. $\nu_L(Y)X - \nu_R(Y)(1-X)$) is the payoff difference between L and R for the type 1 (resp. 2) agent at the threshold X (resp. Y). It is verified that every limit point is an equilibrium; otherwise, the solution path moves away from it. This dynamic process enables us to draw a phase diagram. Two examples of phase diagrams are shown in Figs. 11.3 and 11.4 (see the previous sub-section for the explanation of these figures). Since F and G are Lipschitz continuous, there is a unique solution to the system in $[0,1]^2$ if the initial condition is given. Note that this is a canonical dynamic process. All we need is that the system moves in a (Lipschitz) continuous manner until there is no person who can be better off by switching his standard.

In order to characterize the limiting behavior of the dynamic with the initial state (m,n), let

$$
x^* = \max_{x<m} \{(x,y)|F(x,y) = G(x,y) = 0\}, \tag{11.11}
$$

and

$$
y^* = \max_{y<n} \{(x,y)|F(x,y) = G(x,y) = 0\}, \tag{11.12}
$$

Note that $F(0,0) = G(0,0) = 0$ holds, while $F(m,n) = G(m,n) = 0$ does not hold, and therefore, both x^* and y^* exist. Moreover, since the loci of $F(x,y) = 0$ and $G(x,y) = 0$ are upward sloping, x^* and y^* are uniquely defined, and $F(x^*,y^*) = G(x^*,y^*) = 0$ holds.

Next, it follows from the next lemma found in Smith (1988) that this system is a *monotone system*, i.e., for all initial conditions z_0 and w_0 with $z_0 \geq w_0$ implies $z(t) \geq w(t)$ for all $t > 0$ since

$$
\frac{\partial F}{\partial Y} = f_2[\mu_L(X) + \mu_R(X)] \geq 0,
$$

and

$$
\frac{\partial G}{\partial X} = g_2[\nu_L(Y) + \nu_R(Y)] \geq 0,
$$

where f_2 (resp. g_2) is the partial derivative of f (resp. g) with respect to the second argument.

Lemma 11.1 (Smith 1988, p. 91) *The system given by* (11.10) *is a monotone system if and only if the off-diagonal elements of the Jacobian of* (11.10) *are nonnegative, i.e.,* $\partial F/\partial Y \geq 0$ *and* $\partial G/\partial X \geq 0$.

It also follows from another lemma found in Smith (1988) that in a bounded monotone system, the solution converges to an equilibrium point.

Lemma 11.2 (Smith 1988, p. 94) *Consider a solution* $(X(\cdot), Y(\cdot))$ *in a bounded monotone system. If* $X(\tau) < X(0)$ *and* $Y(\tau) < Y(0)$ *hold for some* $\tau > 0$, *then* $(X(t), Y(t))$ *converges to an equilibrium as t goes to infinity.*

We now have the following proposition.

Proposition 11.3 *The unique equilibrium that is accessible from* (m, n) *is* (x^*, y^*) *given by* (11.11) *and* (11.12).

Proof Consider $K = [x^*, m] \times [y^*, n]$. Since the present system is monotone, $F(x^*, y^*) = G(x^*, y^*) = 0$, $F(m, n) < 0$, and $G(m, n) < 0$ imply $F(x^*, \cdot) \geq 0$, $F(m, \cdot) \leq 0, G(\cdot, y^*) \geq 0$, and $G(\cdot, n) \leq 0$ on K. Therefore, the restriction of (11.10) on K is a self-contained system, i.e., the path would never go out of K so that we can consider it as a bounded monotone system on K.

Suppose that (m, n) is the initial condition. Since $F(m, n) < 0$ and $G(m, n) < 0$ hold, there exists $\tau > 0$ such that $X(\tau) < m$ and $Y(\tau) < n$ hold. Thus, applying Lemma 11.2, we establish that the system converges to an equilibrium. By definition, the only equilibrium in K is (x^*, y^*). Hence, the system converges to (x^*, y^*). $\square$

An immediate corollary of this proposition is that it is always the case that upon lifting the barrier, members of the smaller community assimilate into the larger community. Also, note that the equilibrium accessible from (m, n) does not depend upon the relative speed of adjustment, α and β. Whether the system reaches a completely assimilated equilibrium or a partially assimilated one depends upon μ_R and ν_R. In Fig. 11.3, completely assimilated equilibrium $(0, 0)$ is accessible from (m, n). On the other hand, in Fig. 11.4, the uniquely accessible equilibrium from (m, n) is the partially assimilated equilibrium (x^*, y^*).

Proposition 11.3 allows us to perform some comparative statics. We compare two societies in terms of the degree of assimilation. Consider two societies in both of which (m, n) represents community A as well as the initial point. One society has distributions μ and ν, and the other has μ' and ν'. Let (x, y) be the equilibrium accessible from (m, n) in the first society, and (x', y') be the equilibrium accessible from (m, n) in the second society. If $\mu_R(i) < \mu'_R(i)$ holds for all $i \in [0, m)$ and $\nu_R(i) < \nu'_R(i)$ holds for all $i \in [0, n)$, then we have

$$x \geq x' \text{ and } y \geq y',$$

where strict inequalities hold whenever x' and y' are strictly positive. In other words, if agents in community A of one society dislike R more than those in community A of the other society, then the number of people assimilating into community B in the first society is no more than that in the second society. To see this point, note that *F* shifts upward and *G* shifts downward, which makes the intersection close to the origin.

11.4.2 Incomparable Community Sizes

If neither community is larger than the other in both types of agents (e.g., if $m <$ $1/2 < n$), there may be different equilibria accessible from (m, n) depending upon the relative speed of adjustment between types 1 and 2. Figure 11.5 shows such a possibility. In this case, we cannot determine which community will gain more after the barrier is lifted. If, for example, the speed of adjustment of community A is sufficiently slower than that of community B, then community A will benefit more from free trade since agents in community B adjust to L before those in A adjust to R.

11.4.3 Different Initial Conditions

If we take a different distribution under autarky as the initial condition, then the result would change. The easiest and trivial cases are the ones in which the two communities adopt the same standard from the beginning. These cases are reduced to the ones without multiple standards, and no issue of assimilation would arise.

Another less obvious and potentially interesting case is when agents of each community coordinate upon their less preferred standard, i.e., standard R is taken in community A, while standard L is taken in community B. In this case, we cannot use the technique developed in the main part of the analysis since $i < j$ no longer implies that agent i takes R only if agent j takes R. Indeed, the system becomes

Fig. 11.5 Phase diagram: incomparable community sizes

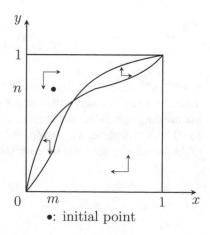

•: initial point

four dimensional, and without some simplification, we cannot analyze its general properties. One simplification is to make the two types symmetric, so that the dynamic of type 1 agents is the same as that of type 2 agents, which we assume throughout this subsection.

For this purpose, we assume that $m = n$ holds, and that $\mu_R(i) = \nu_R(i)$ and $\mu_L(i) = \nu_L(i)$ hold for all $i \in (0, 1]$. Moreover, the adjustment speed is assumed to be the same between types 1 and 2. We still assume, however, that in each community, agent i takes R only if agent j takes R for all i and j with $i < j$. This assumption enables us to characterize the system by two thresholds, z_A and z_B, where for $k = A, B, i < z_k$ (respectively $i > z_k$) implies that agent i of both types of community k takes standard L (respectively R). The system is bounded by $[0, m] \times [m, 1]$. The initial condition $(z_A(0), z_B(0))$ is equal to $(0, 1)$.

11.5 Welfare Implications

The previous section has established that there is the unique equilibrium that is accessible from the autarky equilibrium (m, n) provided that $m, n < 1/2$ holds, which we assume throughout this section. This enables us to turn to the welfare implications of our accounting for the inherent preferences over a set of standards when the two communities are integrated. In the sequel, therefore, we focus on the equilibrium that is accessible from the initial condition (m, n).

We use a simple welfare function in which the utility of each member in the community is given equal weight. Thus, we define total welfare W for community A to be the sum of the total welfare of community A type 1 agents, W_1, and the total welfare of the type 2 agents, W_2. If a pair of thresholds (x, y) satisfies $x < m$ and $y < n$ as in the accessible equilibrium we focus on, they are calculated as follows:

$$W_1 = \frac{xy}{2} + \int_x^m \mu_R(i) \frac{1-y}{2} \, di \tag{11.13}$$

and

$$W_2 = \frac{xy}{2} + \int_y^n \mu_R(i) \frac{1-x}{2} \, di \tag{11.14}$$

where, for example, the first term of (11.13) is the sum of expected payoffs of non-assimilating agents of type 1, and the second term is the sum of expected payoffs of assimilating agents. Note that an assimilating agent i of type 1 obtains the payoff of $\mu_R(i)$ with probability $(1 - y)/2$, and so forth.

In the autarky case, the total welfare of community A is equal to

$$W^0 = \frac{2mn}{m + n}.$$

On the other hand, if $\mu_R(i) = \nu_R(i) = 1$ for all i, i.e., if agents in community A have no inherently preferred standard, then it is verified that integration leads to a completely assimilated equilibrium, $(0, 0)$, and the total welfare of community A is given by

$$\overline{W} = \int_0^m \frac{1}{2} di + \int_0^n \frac{1}{2} di = \frac{1}{2}(m + n). \tag{11.15}$$

The welfare difference between autarky and free trade with no inherently preferred standard is

$$\overline{W} - W^0 = \frac{(m - n)^2}{2(m + n)} \geq 0,$$

where the strict inequality holds whenever $m \neq n$, i.e., each community has a comparative advantage in one good.

Under our assumption, agents in community A typically prefer standard L to standard R. Therefore, in the accessible equilibrium, the total welfare W^* is less than $\overline{W}$. Thus, gains (or losses) from integration can be expressed as

$$W^* - W^0 = (W^* - \overline{W}) + (\overline{W} - W^0),$$

where the first bracket of the right-hand side is the losses caused by switching to a less preferable standard, and the second bracket corresponds to gains from trade. Note that the first bracket is always negative regardless of whether the accessible equilibrium is a completely assimilated or partially assimilated one. It is the relative size of these two effects that determines the overall welfare effect of integration.

To further examine the welfare consequences of integration, we turn to the following explicit example.

Example 1 (**Uniform distribution**) Let $\mu_R(i)$ and $\nu_R(i)$ be distributed uniformly on $(0, 1)$, i.e., they are given by (11.3) and (11.4), respectively. We know from Sect. 11.3 that the stable equilibrium accessible from our initial (autarky) condition (m, n) is the completely assimilated equilibrium, $(0, 0)$. From (11.15), total welfare for the type 1 members of community A at the equilibrium point is given by

$$W_1 = \frac{1}{2m} \int_0^m i \, di = \frac{m}{4}.$$

Similarly, total welfare for the type 2 members of community A is

$$W_2 = \frac{1}{2n} \int_0^n i \, di = \frac{n}{4}.$$

Therefore the total welfare W^* is given by

$$W^* = \frac{m + n}{4}.$$

The amount of welfare change is, therefore, given by

$$
\begin{aligned}
W^* - W^0 &= (W^* - \overline{W}) + (\overline{W} - W^0) \\
&= -\frac{m+n}{4} + \frac{(m-n)^2}{2(m+n)} \\
&= \frac{1}{4(m+n)} \left[(m-n)^2 - 8mn \right].
\end{aligned}
\tag{11.16}
$$

Expression (11.16) is negative if the relative size of m and n is approximately between 0.1 and 10. If comparative advantage is not too strong one way or the other, total welfare decreases as the result of integration. Note that in this case, all agents in community B are better off, and the total welfare is increased by $\frac{(m-n)^2}{2(2-m-n)}$.

11.6 Welfare Under Increasing-Returns-to-Scale Matching Technology

In the last section, we assumed that the matching technology exhibits constant returns to scale, i.e., no matter what the size of a community may be, the probability of an agent's matching with another is always one. This implies that gains from integration are limited to the standard gains from trade based on comparative advantages. When two communities are integrated, however, it is often the case that trade opportunities increase, i.e., the matching technology exhibits increasing returns to scale. In such a case, we have to modify our welfare analysis, including gains from expanding opportunities as a positive effect of integration.

We assume that the probability of matching is proportional to the size of the group, i.e., in the autarky case, the probability that an agent in community A (resp. B) is matched with another agent is $(m+n)/2$ (resp. $(2-m-n)/2$). If the two communities are integrated, the probability of an agent's being matched with another is increased to one. With this additional benefit from integration, it is no longer true that the welfare decreases even if the degree of comparative advantage is small. Indeed, even in the case of $m = n$, it is verified that if $\mu_R(i) = i/m$ and $\nu_R(i) = i/n$ as in Example 1, the welfare increases as the result of integration since the welfare in the autarky case is now $mn/2$, which is less than W^* for all m and n. This by no means implies that integration always leads to an increase in the welfare of community A. To see this point, we turn to the following specific examples. In particular, Example 3 shows that in the presence of three standards, it is possible that both communities are worse off after integration in spite of increasing returns to scale in the matching technology.

Example 2 (**Two subgroups**) We study the example in which all members of community A belong to one of two subgroups in terms of their inherent preferences: a fraction η of the type 1 agents in A have cost $\mu_R(i) = \bar{\mu}_R$ while the remaining $1 - \eta$ of this population have cost $\mu_R(i) = \underline{\mu}_R$, where $\bar{\mu}_R > \underline{\mu}_R$; similarly, a fraction η of the type 2 agents in A have cost $\nu_R(i) = \bar{\nu}_R$ while the remaining agents of this type have cost $\nu_R(i) = \underline{\nu}_R$, where $\bar{\nu}_R > \underline{\nu}_R$.

We assume that $\underline{\mu}_R$, $\bar{\mu}_R$, $\underline{\nu}_R$, and $\bar{\nu}_R$ are such that, in equilibrium, players with costs equal to either $\bar{\mu}_R$ or $\bar{\nu}_R$ will switch to using standard R, while players with costs $\underline{\mu}_R$ or $\underline{\nu}_R$ will continue to use L. Specifically, this means that $\bar{\mu}_R > n/(1-n)$ and $\bar{\nu}_R > m/(1-m)$, while $\underline{\mu}_R < ((1-\eta)n)/(1-(1-\eta)n)$ and $\underline{\nu}_R < ((1-\eta)m)/ (1-(1-\eta)m)$. From these conditions, we see that for a given $\bar{\mu}_R$ and $\underline{\mu}_R$, there is a range of n for which a partial assimilation equilibrium exists. An equivalent statement can be made for $\bar{\nu}_R$ and $\underline{\nu}_R$.

Given this, the total welfare of community A type 1 agents in equilibrium is given by

$$W_1 - \frac{1}{2} \int_0^{(1-\eta)m} (1-\eta)n \, di + \frac{\bar{\mu}_R}{2} \int_{(1-\eta)m}^m [1-(1-\eta)n] \, di,$$

which is equal to

$$W_1 = \frac{1}{2}(1-\eta)^2 mn + \frac{\eta\bar{\mu}_R}{2}[1-(1-\eta)n]m.$$

This expression, while somewhat complicated, is readily interpretable. The first term in the expression represents the utility level of the community members who continue to use L. This mass of members, in the autarky case, would have received welfare level

$$\int_0^{(1-\eta)m} \frac{n}{2} \, di = \frac{1}{2}(1-\eta)mn > \frac{1}{2}(1-\eta)^2 mn.$$

These members have suffered a welfare loss. This is, of course, the direct result of the negative externality imposed upon them when the $\bar{\mu}_R$ and $\bar{\nu}_R$ members of their community switch to R.

On the other hand, for their incentive constraint to have been satisfied, the agents with $\bar{\mu}_R$ and $\bar{\nu}_R$ must have experienced a welfare gain. This is easy to verify. The second term in the above expression represents the new level of welfare that these agents receive. Previously, again referring to the autarky case, they received

$$\int_{(1-\eta)m}^m \frac{n}{2} \, di = \frac{1}{2}\eta mn,$$

which, given our initial restrictions on $\bar{\mu}_R$ and $\bar{\nu}_R$, is strictly less than their new level of welfare.

We now ask whether or not the welfare gain experienced by the agents using R outweighs the welfare loss incurred by those continuing to use L. We will look specifically at the case where $\bar{\mu}_R = \bar{\nu}_R = 1$, since if the inequality holds under this condition, it will certainly hold in the case where $\bar{\mu}_R$ and $\bar{\nu}_R$ are less than 1, as in this case, the welfare gain experienced by the gaining agents is diminished. As before, we look first at the net change in welfare that the type 1 agents experience, and then we can examine, separately, the type 2 agents. If the type 1 agents have incurred a net welfare loss, the following inequality will hold:

$$\eta(1-\eta)mn > \eta m[1-(1-\eta)n] - \eta mn,$$

which gives us the condition

$$n > \frac{1}{3-2\eta}.$$

Given that the relevant inequality for the type 2 agents is symmetric, we can conclude that the following condition will also hold if the type 2 agents experience a net welfare loss:

$$m > \frac{1}{3-2\eta}.$$

Again, these conditions are sufficient, but if $\bar{\mu}_R$ and $\bar{\nu}_R$ are strictly less than one, then weaker conditions will suffice. Either way, these conditions tell us immediately that for ranges of m and n, a net welfare loss may result with the expansion of trade opportunities when the costs of coordination are considered.

Example 3 (**Three standards**) We turn now to consider a situation in which agents may choose from among three behavioral standards, namely L, C and R. We look at a rather specific example. Let the payoff matrix for type 1 agents be that in Table 11.1. We assume that a corresponding matrix applies to type 2 agents. However, as before, we focus our welfare analysis on the type 1 agents since the analysis for the 2 agents is symmetric.

We assume that the communities A and B are equal in size, i.e., $m = n = 1/2$. We retain the assumption that the standard L is the most preferred standard by members of community A while R is most preferred by members of B. Thus, we retain the normalization that for type 1 members of A, $\mu_L(\cdot) = 1$, while for members of B, $\mu_R(\cdot) = 1$. Furthermore, we retain the initial condition at (m, n), i.e., all members of A use L and all members of B use R. Note, however, that x and y as previously defined are no longer sufficient to characterize equilibrium here because of the addition of the third standard.

The welfare results in this case do not rely upon there being heterogeneity among agents with respect to μ_L and μ_R, although the results do hold for appropriate parameter values when heterogeneity is present. Thus, for ease of enumeration, we will

Table 11.1 Payoff table for the case with three standards

	L	C	R
L	μ_L	0	0
C	γ	μ_C	γ
R	0	0	μ_R

assume that agents within a community are homogeneous in this regard. Furthermore, we can assume that all agents in both communities earn the payoff γ when using C and trading with someone using L or R, while earning the payoff μ_C from using C and trading with someone using C.

We assume that $\gamma > \max\{n, 1 - n\}$ and that if agents from A and B switch to standard C from their preferred standards, they do so at the same rate. We use the remainder of this section to show that if

$$\gamma > \frac{1}{2} > \mu_C > \frac{1}{2}\gamma - \frac{1}{4}, \tag{11.17}$$

then the lifting of a trade barrier between A and B leads to a welfare loss for every individual in both A and B.

Given our assumptions, the stable equilibrium accessible from our initial condition is the equilibrium in which all members of both A and B choose standard C. To see this, we first consider the decision of an agent at the initial point when the barrier is lifted. If the agent is a member of A, then taking L offers an expected payoff of $\frac{1}{2}n = \frac{1}{4}$, taking C offers $\frac{1}{2}\gamma$, and taking R offers $\frac{1}{2}\mu_R(1 - n) = \frac{1}{4}\mu_R$. The agent's best/better response is clearly to choose C, given our assumption regarding the value of γ. The same argument holds for members of B. Thus, we expect some agents to switch to C.

Now, since agents from A and B switch to C at the same rate, we can say that at some fixed point in time, a fraction c of the agents in both groups are using C. Thus, in evaluating his options, an agent in A sees that his expected payoff equals $\frac{1}{2}(\frac{1}{2} - c)$ if he chooses L, $\frac{1}{2}[\gamma(1 - 2c) + \mu_C(2c)]$ if he chooses C, and $\frac{1}{2}\mu_R(\frac{1}{2} - c)$ if he chooses R. Since choosing R is clearly a dominated strategy, we need only assess the comparison between his choosing L and C. In doing so, we find that if the following equation holds, then an agent will still prefer C to L if the following inequality holds:

$$\gamma(1 - 2c) + \mu_C(2c) - \left(\frac{1}{2} - c\right) > 0.$$

This equation will hold for all $c \in [0, 1]$ if it holds for $c = 1$. Thus, we find that if

$$\mu_C > \frac{1}{2}\gamma - \frac{1}{4},$$

then in equilibrium, all agents will stay with the choice C. In equilibrium, the payoff expected by every agent equals $\frac{1}{2}\mu_C$. If $\mu_C < \frac{1}{2} = n = 1 - n$, then the expected payoff to every agent is lower than it was at the initial condition. Therefore, we say that all agents in both communities experience a welfare loss upon the lifting of the trade barrier between the communities.

11.7   Remarks

We have highlighted the importance of explicitly considering the need for coordination in interactions when modeling economic behaviors where such coordination is required. We have shown that when we account for the costs of such coordination, there are cases in which total welfare of a minority community decreases when a trade barrier between the two communities is lifted. In addition, we offered an example that illustrates that in a situation where no dominant culture exists, every member of both communities may ultimately be worse off upon the lifting of a trade barrier.

A few remarks are in order. First, the present analysis has not considered important inter-generational issues that are pertinent in any discussion of assimilation. It is often argued that one of the most serious problems associated with assimilation is the gap that arises between generations. Parents become alienated from their children and cannot pass on the wisdom they have inherited from generations of people that came before them. Children who wish to assimilate must learn the new culture on their own. They often remain second-class citizens in the new society. This effect may persist, in some cases becoming intensified and while in others, becoming weaker. In cases where this effect becomes larger, the rate of economic growth may be higher for members of a dominant group in society than for those coming from a minority group. In addition to this problem, we have assumed that people make their choices myopically. We have not considered the case in which people take into account future generations when making their own decisions regarding assimilation.

Our next remark is related to our first. We do not presently deal with situations in which discrimination makes it essentially impossible for one group to coordinate with, or assimilate into, another group. This problem arises most commonly in cases when a group has some recognizable traits that cannot be changed, even by choice, such as gender or skin color. In repeated situations, discrimination is sustainable in equilibrium even if the only difference between people is their "names" (Chaps. 13 and 14). In such cases, it may be that members of one community would like to coordinate with the members of another group, but when they take the appropriate behaviors that would seemingly allow them to do so, they effectively end up as a group unto themselves, forced to interact primarily within the newly formed, third group.

Third, if some people can switch between two standards, these people may act as middlemen between the two communities. Examples are international merchants and English-speaking Chinese-Americans in Chinatown. In these cases, while the minority may not lose their identities, the wealth could be concentrated on a handful of middlemen.

Fourth, different situations present different problems. For example, in the case of computer networks, standardization may imply the need for complete coordination. On the other hand, culture cannot be described by a single trait (Cavalli-Sforza and Feldman 1981). Adopting one trait but not another may have effects that we do not capture with the model in this analysis. Thus, we must more carefully examine the contents of such traits when we apply our analysis to specific problems. One typical question that must be addressed in this vein is the question of which traits can be changed and at what cost. As can be seen, the model presented in the present analysis is far from a universal one. We only suggest one of many possibilities. Still, it raises an issue that has been ignored in the literature. If the reader realizes that the demand for coordination sometimes offsets gains from trade, half of the goal of the present analysis would be achieved.

Chapter 12
A Model of Man as a Creator
of the World

12.1 Introduction

This chapter proposes a theory of man, wherein man constructs models of the world based on past experiences in social situations.[1] The present theory considers experiences, or chunks of impressions, as primitives instead of an "objective" game, which is assumed to be given in the standard economic theory. Agents construct models of the world based on direct and indirect experiences. Each model comprises a structural part and a factual part. The structural part is represented as a game, while the factual part is represented as a strategy profile of this game. In constructing a model, an agent might use certain axioms, for example, coherence, according to which the model should be able to explain his or her own experiences; conformity to a solution concept; and minimality with respect to some simplicity measure.[2]

For more than a century, differences in intellectual ability between human beings and other species have been studied extensively (see, e.g., Thorndike 2017 for some earlier works). Many "intelligent" activities, especially those analyzed by Simon (1957), are now known to be shared not only by primates but also by a variety of animals. Many birds and mammals are known to use their intelligence to try to behave satisfactorily, if not optimally, in various situations. They too learn how to hunt, fly, and breed. For example, it is commonly observed that birds bred by humans can neither fly nor breed by themselves.[3] Furthermore, studies with an African grey parrot found that the bird was able to demonstrate numerical competence.[4] Needless

---

[1] This chapter is based on Matsui (2008).

[2] Peirce (1898/1992) called this activity *retroduction* (or abduction), claiming that we had to distinguish this activity from the "standard" induction by which we enlarge our observation from samples to the entire population. I am grateful to Takashi Shimizu for pointing this out to me. See Matsui and Shimizu (2007) for more discussion.

[3] There are numerous reports on the difficulty of animals' returning to the wild. Many programs are designed to teach animals various skills to survive in the wild. See, e.g., Hendron (2000).

[4] Pepperberg (1994) reported that an African gray parrot (*Psittacus erithacus*), Alex, trained to label vocally collections of 1–6 simultaneously presented homogeneous objects, correctly identified,

© Springer Nature Singapore Pte Ltd. 2019
A. Matsui, *Economy and Disability*, Economy and Social Inclusion,
https://doi.org/10.1007/978-981-13-7623-8_12

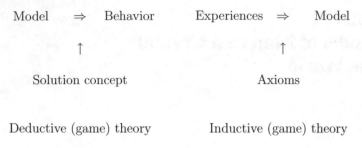

Fig. 12.1 Deductive and inductive theories

to say, humans use their instinct, like other animals, to avoid danger and to react to certain stimuli. Tendencies of such behavior have been extensively studied in psychology and, more recently, behavioral economics in the context of strategic interaction (see, e.g., Camerer 2003).

Nevertheless, humans are distinct from other species with respect to the manner in which they use intelligence. One of the intellectual activities that are often observed in humans, but not in other animals, is the construction of a model of the world that explains their experiences.[5] Focusing on the observation that experiences play a major role in shaping the human mind, the present research proposes a formal game-theoretical framework to study such activities of humans in social situations.

The difference between the standard theory and the present theory is summarized in Fig. 12.1. Standard game theory takes a model, or the structure of a game, as given and applies a solution concept such as Nash equilibrium or a behavior rule to the model in order to derive the strategies/behavior of the players of the game. In the sense that a specific act is induced by a general principle, the present analysis calls this theory the *deductive game theory*.

On the other hand, the present theory takes experiences, or chunks of impressions, as primitives. Based on them, an agent constructs a model. Some axioms are used in constructing a model. In the sense that a general structure of the game is induced based on limited experiences, the present analysis calls this theory *inductive game theory*.

Three remarks are in order. First, these two theories should not be regarded as substitutes; rather, they are complements. In reality, people use both induction and deduction in accumulating knowledge, which eventually affects their behavior. Second, the construction of a model does not have to precede action taking place. Rather,

---

without further training, quantities of targeted subsets in heterogeneous collections. For each test trial Alex was shown different collections of 4 groups of items that varied in 2 colors and 2 object categories (e.g., blue and red keys and trucks) and was asked to label the number of items uniquely defined by the conjunction of 1 color and 1 object category (e.g., "How many blue keys?"). The collections were designed to provide maximal confounds (or distractions). Unfortunately, further tests cannot be conducted since Alex died on September 6, 2007, at the age of 31.

[5] However, it is difficult to reject the hypothesis that animals, too, perform such an intelligent activity of constructing a model of the world, in a broad sense.

experiences typically include those impressions that are obtained through one's own behavior. Third, the present framework takes game theory as a language that agents use to describe the world rather than as a refutable "theory". Although the present way of describing models in the language of game theory is far from being general, it enables us a rigorous study of the situations that agents experience.

Several applications are presented to show the basic workings of the theory. The first application is concerning entry and predation. The failure of Air Do, a Japanese regional airline, illustrates the workings of this theory. The second one is with regard to bullying. Through the activity of bullying in school, children may construct a specific way of viewing a situation. The third application concerns the importance of pioneers.[6]

The idea of the construction of models by agents based on experiences was initiated by Kaneko and Matsui (1999), who examined a specific game called the festival game. Subsequent papers by Kaneko and Kline (2008) and Matsui and Shimizu (2007) are closely related to the present analysis. Kaneko and Kline (2008) proposed the concept of information protocol and demonstrated a correspondence between games expressed in extensive form and in information protocols.

Matsui and Shimizu (2007) confined their attention to the class of repeated games and sought conditions under which an objective game and a subjectively constructed model coincide. The present analysis does not presume the existence of an objective game, and therefore, does not pay attention to the conditions under which agents can reconstruct the objective game from experiences.

The present analysis focuses on induction as the main inference rule. In this regard, the present work shares a common thread with a sequence of works by Gilboa and Schmeidler (1995, 2001), Fudenberg and Levine (1993), and Battigalli and Siniscalchi (2002). However, a critical difference is on the decision-making process, whereas the present analysis focuses on man's creation of models of the world. Referring to Peirce (1898/1992), Matsui and Shimizu (2007) argued that the kind of activity analyzed in the present analysis should preferably be called *retroduction*, or the *inference to the best explanation*, as subsequent researchers have called it.

The rest of the chapter is organized as follows. Section 12.2 presents the basic framework of the theory. Section 12.3 describes some applications. Section 12.4 presents a motivational background, referring to Hume and his influence on Einstein. Section 12.5 concludes the chapter.

---

[6]Another application, which will be discussed in Chaps. 13 and 14, concerns discrimination and prejudices as discussed in Kaneko and Matsui (1999). Instead of the standard argument on this subject (Becker 1957), i.e., that prejudices lead to discrimination, the present framework allows an argument that the fact of segregation gives rise to prejudices against the segregated.

12.2 Inductive Construction of Models

12.2.1 Impressions and Experiences

Agents accumulate experiences, each of which constitutes a chunk of impressions sensed and felt by the agent. Experiences and impressions are primitives of the current framework. A generic agent is denoted by i. A generic experience is given by $\varepsilon_i = (\varepsilon_{i1}, \ldots, \varepsilon_{iK})$. ε denotes the set of all possible experiences.

12.2.2 Models

We use the standard representation of games as models of the world that the agents construct. Based on the set of experiences, each agent constructs a *model*, which represents his or her understanding of the situation in question. A *model* is generically given by

$$m = \langle \Gamma, \sigma \rangle,$$

where Γ is the *structural* part of the model, which is a (modified stochastic) game, and σ is the *factual* part of the model, which is a strategy profile of the game Γ. Let $\mathcal{M}$ be the set of all such models.

Some models are of special interest. Here, we mention two classes of such models. The first class is that of repeated game models. For phenomena ranging from the sunrise to daily activities, a person often views the situations he or she faces as if it would repeat indefinitely. A repeated game model captures this view in a simple manner. The second class is that of random matching models, which are often constructed to study a society where each agent faces the same situation over a period of time without getting involved in intertemporal strategic considerations.

12.2.3 Axioms

Axioms are the criteria that agents use to construct models of the world. There is no axiom that *ought* to be used *a priori*. Axioms themselves may be in flux in the human mind, similar to a researcher adopting different axioms from time to time. However, there are some that are considered plausible. The first of such axioms is *coherence*, which requires that a model be able to explain one's experiences.

Axiom 1 (*Coherence*) Given an experience ε_i, $\langle \Gamma, \sigma \rangle$ is *coherent* with ε_i if ε_i is induced by σ with a positive probability in Γ.

In addition, we may add another condition to consider the notion of statistical coherence. Let a random variable induced by a model $\langle \Gamma, \sigma \rangle$ be denoted by $\varepsilon_{i\langle \Gamma, \sigma \rangle}$. Then, we have the following axiom.

Axiom 2 (*Statistical Coherence*) Given a set T of statistical tests and an experience ε_i, a model is *statistically coherent* if, in addition to Axiom 1, the null hypothesis that ε_i is induced by $\varepsilon_{i\langle \Gamma, \sigma \rangle}$ is not rejected by either of these tests.

We do not define this axiom more rigorously because the way it is defined depends upon the set of statistical tests to be used. We do not use statistical coherence in the subsequent applications.

A *solution concept* is a correspondence ψ that maps a stochastic game to a set of strategy profiles (possibly empty for some games). It is defined without referring to experiences.

Axiom 3 (*Conformity*) Given a solution concept ψ, a model $m = \langle \Gamma, \sigma \rangle$ *conforms* to the behavior rule ψ if $\sigma \in \psi(\Gamma)$.

An example of a solution concept is Nash equilibrium. Other examples include solution by backward induction and behavior rules used in, say, learning theories.

Axiom 4 (*Uniqueness of Outcome/Solution*) Given a model $m = \langle \Gamma, \sigma \rangle$, a solution concept ψ induces a *unique outcome* if all $\tilde{\sigma}$'s in $\psi(\Gamma)$ induce the same stochastic process of outcome. ψ induces the *unique solution* if $\psi(\Gamma) = \{\sigma\}$.

The following two axioms are controversial in the philosophy of science. Nonetheless, in reality, both scientists and laymen tend to use them. The first is the principle of simplicity, and the second is that of observability.

Given $M \subset \mathcal{M}$, let $\geq_M$ denote a binary relation on M. We write $m >_M m'$ if $m \geq_M m'$ holds, but not $m' \geq_M m$.

Axiom 5 (*Minimality*) Given $M \subset \mathcal{M}$ and a binary relation $\geq_M$ over M, a model m is said to be *minimal* with respect to $\geq_M$ on M if there exists no $m' \in M$ satisfying $m >_M m'$.

Different agents may use different binary relations. A confused agent may have an intransitive binary relation; however, we may assume that $\geq_M$ is a preorder for most cases.[7]

The next axiom is a principle of observability, according to which one tends to choose a model that explains situations, especially payoff functions, only by observables. Given a model $m = \langle \Gamma, \sigma \rangle$, an *observable variable* is a function X that maps each outcome of Γ to something that can be observed by agents.

Axiom 6 (*Observability*) Given a sequence $(\varepsilon_{i1}, \ldots, \varepsilon_{iK})$ of experiences, a model $m = \langle \Gamma, \sigma \rangle$ satisfies the *principle of observability* if payoffs and strategies are the functions of observable variables.

---

[7] A binary relation $\geq_M$ is a *preorder* if it satisfies reflexivity and transitivity, i.e., $[\forall x \in M (x \geq_M x)]$ and $[\forall x, y, z \in M (x \geq_M y \wedge y \geq_M z \Rightarrow x \geq_M z)]$, respectively.

12.2.4   Prior Beliefs

Prior to the construction of a model based on experiences, agents may have held certain beliefs concerning the situation. At this point, we disregard the source of these beliefs, for instance, whether they were derived from pure reasoning or from prior experiences. These beliefs may take various forms. A possible representation of such a belief, at least from the viewpoint of researchers, is to restrict the possible models to a certain class. Let $M \subset \mathcal{M}$ be a subset of games in extensive form. An agent's prior beliefs can be represented by such an M.

Note that in applications, we often assume agents to have some prior beliefs, which tend to be inductively obtained from prior experiences. Therefore, one should regard the term "prior" beliefs not as *a priori* beliefs in a philosophical sense, but as assumptions for the sake of analysis.

12.3   Applications

12.3.1   Predation

> Experience is a dear teacher, but fools will learn at no other.
>
> —Benjamin Franklin

In 1998, after raising money from the general public, Air Do entered the Japanese domestic airline market after the deregulation of the airline industry in the 1990s. Air Do was one of Japan's first low-fare airlines, operating between Chitose, Hokkaido and Haneda, Tokyo. Initially called "*Do-min no Tsubasa*"(the wing of Hokkaido-residents), it provided its passengers with low-fare flights between Tokyo and Hokkaido. It competed with Japan's major domestic carriers [All Nippon Airways (ANA), Japan Airlines (JAL), and Japan Air System], which lowered their fares to Air Do's level without extensively compromising on corporate profits. After two years of incurring losses and despite continuous financial support from the local government of Hokkaido, Air Do went bankrupt, retired all its stocks, and entered into a code-sharing agreement with ANA. Not only did Air Do lose its money, it also lost its dream of becoming "the wing of Hokkaido-residents," adopting the same general fare structure as the major airlines. ANA seems to have emerged as the winner of this predation game because it acquired Air Do as a low-cost airline. Indeed, ANA has had Air Do expand its routes from only one (Haneda-Chitose) to four (Haneda-Chitose, Haneda-Asahikawa, Haneda-Hakodate, and Haneda-Memanbetsu).

To understand this situation, suppose that a potential entrant E considers whether or not to enter a market monopolized by an incumbent I before deregulation. From some other markets of similar characteristics, E learns that once an entrant enters, an incumbent often acquiesces, and the two firms share the market accordingly.

Formally, assume that an entrant E's (indirect) experiences concerning, for instance, the US airline market, are

$$\varepsilon_{E'1} = (Regulation, (E' : \{not\}), (E' : not), (I' : \text{``4''}), (E' : \text{``0''})),$$

before deregulation, where $(E' : \{not\})$ implies that E had the only option of "not enter"; $(E : not)$ implies that E' chose "not enter"; $(I' : \text{``4''})$ implies that the incumbent $I'$ obtained the payoff of 4; and $(E' : \text{``0''})$ implies that E' obtained zero, and

$$\varepsilon_{E'2} = (Deregulation, (E' : \{not, enter\}), (E' : enter), (I' : \{p_H, p_L\}), (I' : p_H),$$
$$(E' : \text{``1''}), (I' : \text{``2''})),$$

after deregulation, where $(E' : \{not, enter\})$ implies that E' has the two options, "enter" and "not enter"; $(E' : enter)$ implies that E' chose to "enter"; $(I' : \{p_H, p_L\})$ implies that I' had options of p_H and p_L; $(I' : p_H)$ implies that I' took p_H; $(I' : \text{``2''})$ implies that $I'$ obtained the payoff of 2; and $(E' : \text{``1''})$ implies that E' obtained 1. In this description, p_H (resp. p_L) denotes a high (resp. low) price. Having observed them, E constructs a model.

Let us consider games in extensive form to represent this situation.

The game shown in Fig. 12.2 is coherent with E's experiences irrespective of the values v and w.

Suppose that ψ complies with backward induction. Then, the conformity axiom requires $2 > v$. If the simplicity relation is such that the less the number of edges, the simpler is the model, then this model is a minimal one (unique up to the values of v and w) among the coherent models.

On the other hand, a model created by I may be represented as in Fig. 12.3. Here, we assume that after I takes the option of *fight*, E has an option of *exit* or, alternatively, *stay*.

In this game, E has to be cautious and, in fact, refrain from entry since it should expect a negative profit if it believes that ψ complies with backward induction.

In the case of Air Do, it could not bear the loss caused by the predatory pricing of the two incumbents, ANA and JAL, and exited the market, or to be precise, retired its capital and reached a code-share agreement with ANA. Air Do failed to construct a correct model to evaluate the situation it faced.

Fig. 12.2 Predation game

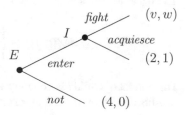

Fig. 12.3 Predation game
with an exit option

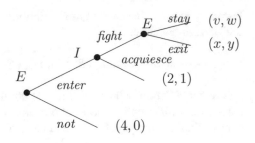

12.3.2 Bullying

> You will probably be bullied wherever you may go unless you have some fighting spirit.
>
> —Shintaro Ishihara[8]

Suppose that there are four children, A, B, C, and D, which is already part of prior knowledge. Also, every child has prior knowledge that each time two children meet in pairs, they simultaneously decide whether to play in a friendly (F) or unfriendly (U) manner. Suppose now that these children have observed that each time two of A, B, and C have met in pairs, they took F and looked happy, while when they met D and formed a pair, they played U, while D played both F and U from time to time, and the two looked unhappy. In addition to these impressions, the children observed various attributes of each other, such as color, height, body shape, face, and others. Let ε_j denote such an experience of Child j ($j = A, B, C, D$).

There are numerous models that are coherent with the above experience even if we restrict our attention to random matching models. Here, we consider two classes. The first one assumes some intrinsic differences between the children, whereas the second does not. In the both models, Nature randomly determines a pair of children to be matched. After two children are matched, they play a simultaneous move game where both of them have two available acts F (Friendly) and U (Unfriendly), which the children are also aware of.

In the first model of Child $i = A, B, C$, a payoff function can take the following form:

$$u_i(a_i, a_j; j) = \begin{cases} 1 & \text{if } a_i = a_j = F \text{ and } j \neq D, \\ 0 & \text{otherwise, } (i = A, B, C). \end{cases} \tag{12.1}$$

The behavior rules of A, B, and C are given by

$$\sigma_i(\{i, j\}) = \begin{cases} F & \text{if } j \neq D, \\ U & \text{if } j = D, (i = A, B, C). \end{cases}$$

This model of Child i is coherent with ε_i, and their strategy profile constitutes a Nash equilibrium. Moreover, under some "reasonable" criteria of minimality such as the

---

[8] A remark at press conference on Nov. 10, 2006; translated by the author.

one that counts "complexity" by the numbers of acts and payoff values, this model becomes minimal.

If it happens to be the case that D is taller than the other three children, then children may construct a model in such a way that they do not enjoy playing with a tall child. To construct such a model, suppose that h_j $(j = A, B, C, D)$ is the height of Child j, and that $h_j < \bar{h}$ for $j \neq D$, while $h_D > \bar{h}$ where $\bar{h}$ is a threshold value. In this case, we have

$$u_i(a_i, a_j; h_j) = \begin{cases} 1 & \text{if } a_i = a_j = F \text{ and } h_j < \bar{h}, \\ 0 & \text{otherwise}, (i = A, B, C). \end{cases} \quad (12.2)$$

in place of (12.1). The purpose of this analysis is to show that any attribute can be a reason for bullying.

In the second model, each child obtains one as a payoff if both choose F, and zero otherwise.

$$u_i(a_i, a_j) = \begin{cases} 1 & \text{if } a_i = a_j = F, \\ 0 & \text{otherwise}, (i = A, B, C, D). \end{cases} \quad (12.3)$$

Each child plays a repeated game strategy according to which they determine a "target" and play U whenever a child meets the target child, and they continue to do so until someone takes F against the target, after which the child who chose F now becomes a new target. This strategy profile is a subgame perfect equilibrium of the constructed repeated game.

12.3.3 Pioneers

> I think the importance of being a pioneer is that you have to be successful, ... Being successful leads to the next player, and the next player and so on.
>
> —Don Nomura, the agent of Hideo Nomo[9]

In 1995, Hideo Nomo, a Japanese pitcher, signed a contract with the Los Angeles Dodgers after a contract dispute with Kintetsu Buffaloes, a professional baseball team in Japan. He was only the second Japanese baseball player to make a Major League debut, only after the nearly forgotten Masanori Murakami. Nomo's games were regularly broadcast in Japan. Unlike Murakami, Nomo exceeded the expectations of the Japanese media and fans. His success inspired many baseball star players like Ichiro and Hiroki Kuroda to move to the United States, too. Before Nomo, neither Japanese player nor club team had ever even dreamed of succeeding in Major League Baseball (MLB). Nobody ever predicted before 1995 that Japanese players

---

[9]Quoted in the article "Wally Yonamine" by Rob Smaal in the English edition of Asahi.com, January 2, 2007.

could compete with MLB players. A transfer to MLB was not even in their scope. After 1995, a door to MLB unexpectedly opened to Japanese players all the sudden.

Pioneering works have one thing in common. All of them change the scope of people. In fact, this is almost the definition of a "pioneer." After having observed numerous instances of Japanese players' successes and failures with respect to playing in Japan, and, with respect to playing in the USA, only one forgotten instance of failure and no success, it is not difficult to imagine that people construct a model, wherein playing in the USA is not even an option.

12.4   Induction and the Science of Man

Game theory should pay more attention to induction than it does now, though we do not have to deny the importance of deduction. Hume considered induction as an activity of the human brain more fundamental than deduction, in obtaining new knowledge. We now turn to the theory of Hume to note the importance of induction in developing the science of man.

12.4.1   Hume

Hume (1711–1776) tried to establish the science of man that not only corresponds to, but also serves a basis of the science of matter established by Newton, paying attention to the activity of induction. In this regard, Hume (1739/1984) wrote the following:

> Even Mathematics, Natural Philosophy, and Natural Religion, are in some measure dependent on the science of MAN; since they lie under the cognizance of men, and are judged of by their powers and faculties... If therefore the sciences of Mathematics, Natural Philosophy, and Natural Religion, have such dependence on the knowledge of man, what may be expected in the other sciences, whose connexion with human nature is more close and intimate? (A Treatise of Human Nature; Introduction)

Let us briefly look at his theory in an intuitive manner.

Say that we ask a toddler "Do you want some fruit?"

She replies "No," and pokes a finger at an orange.

Before laughing at the toddler, let us also think about the fruit. Can we imagine something that is not an orange or an apple, but that looks like a "fruit"? One may think of an orange, and another may think of a fruit basket. There is no such thing as a "fruit" that is neither orange nor apple, etc. What exist are oranges and apples, and "fruit" is a concept that holds some of their common characteristics.

One might think this is because "fruit" is an abstract concept. In fact, "orange" and "apple" are similar concepts. Let us now consider an apple one is about to eat.

A toddler points at a green apple and asks, "What is it?" You answer, "An apple." But she replies, "No. Apples are red."

After eating the apple, she understands. To the toddler, an "apple" is a totality of "redness," "roundness," and "bitter sweetness."

Hume investigated this problem further and applied the same argument to "I." Elaborating the term "I feel," I encounter the feeling of "coldness" or "warmness." Likewise, when "I think," what exists is the very thinking such as "Hume established the science of man." "I" do not exist independent of such thinking and feeling; rather, "I" am the totality of them.

Hume (1739/1984) divides human perceptions into two distinct types, impressions and ideas:

> Those perceptions, which enter with most force and violence, we may name *impressions*; and under this name I comprehend all our sensations, passions and emotions, as they make their first appearance in the soul. By *ideas* I mean the faint images of these in thinking and reasoning. (Hume 1739/1984; BOOK I, PART I, Sect. I)

According to Hume, in our thinking and reasoning, we associate one idea with another in three ways, resemblance, contiguity in time or place, and cause and effect. What made Hume original is his argument on cause and effect, to which we now turn.

There is a Japanese saying that goes "No fire, no smoke," which implies that causes precede effects. It is this causal relation that connects the ideas of "fire" and "smoke." We think that there are causes to many social problems and try to solve them by finding their causes.

However, Hume argued that we cannot sense the causal relation. After repeated observations of fire and smoke, we come to conclude the relation "fire $\Rightarrow$ smoke."

Of course, we do not always term what comes first as a cause, and what happens next as an effect. It sounds silly to claim that a cock's crowing causes the sunrise even if one always hears it before the sunrise. However, if one is asked why they believe the causal relation "fire $\Rightarrow$ smoke," but not "cock's crow $\Rightarrow$ sunrise," then one has to resort to another experience like, "The sun rises even without a cock's crowing." Hume (1739/1984) claims, "that all our reasoning concerning causes and effects are driv'd from nothing but custom; and that belief is more properly an act of the sensitive, than of the cogitative part of our natures (BOOK I, PART IV, Sect. I)."

Before Hume, the inference of the form "cause $\Rightarrow$ effect" was often confused with the logical inference of the form "premise $\rightarrow$ consequence." The two premises "Socrates is a man" and "Man is mortal" lead to the consequence "Socrates is mortal." This inference is deduction, which gives us no new knowledge since this consequence is always true if the above two premises are true. To establish a new piece of knowledge such as "fire $\Rightarrow$ smoke," one needs induction, of which logical necessity was denied by Hume.

Since our knowledge is based upon causes and effects, this argument leads to skepticism. Even seemingly obvious claims like "the sun will rise to-morrow" and "all men must dye (*die*)" are derived from nothing but custom. This argument, which sounds as if it were a mere philosophical discourse, has a new meaning in the current world when we confront a situation in which beliefs like "the forest has been and will be there forever" are negated.

Unlike subsequent philosophers, who tried to overcome the Humean curse of skepticism, he himself did not adhere to it; rather, he used it as leverage to counter a variety of preconceptions that people of his age frequently possessed.

Even the science of matter could not escape from the Humean argument. However, Hume's theory became a catalyst rather than an obstacle to the development of the science of matter. Later, Albert Einstein wrote to Werner Heisenberg, "It is the theory that decides what we can observe." Einstein wrote that Hume had been there to assist the birth of the special theory of relativity. We now turn to his relationship with Hume.

12.4.2 *Einstein*

Around the turn of the nineteenth century, a problem that puzzled Einstein and other physicists was the paradox of light and simultaneity. He recollected the moment of the discovery of his idea of special relativity:

> Today everyone knows, of course, that all attempts to clarify this paradox [that induced the special theory of relativity] satisfactorily were condemned to failure as long as the axiom of the absolute character of time, or of simultaneity, was rooted unrecognized in the unconscious. To recognize clearly this axiom and its arbitrary character already implies the essentials of the solution of the problem. The type of critical reasoning required for the discovery of this central point was decisively furthered, in my case, especially by the reading of David Hume's and Ernst Mach's philosophical writings.
>
> —Albert Einstein (Schilpp 1979, p. 51)

To understand his theory in an intuitive manner, consider a train running at a constant speed with its windows shut. Suppose that two baseball pitchers standing at the two ends of the train simultaneously throw balls at a target in the center of the train in the same manner. Which ball hits the target first? The answer is "the two balls hit it simultaneously." At the same time, to an observer sitting at the center of the train, the speeds of the two balls appear to be the same.

Next, consider an observer standing beside the track. The ball thrown in the direction in which the train is moving appears faster than the speed of the train, while the ball thrown in the opposite direction appears slower than the train's speed. The relative speed of the two balls appears different to the two observers.

Let us now replace the balls with light beams (and also imagine an inter-galaxy train). The story changes in this case. Light is known to move at the same speed irrespective of the relative movement of observers. Nothing has changed essentially for the observer inside the train. Light beams emitted simultaneously from each end hit the target at the same time. On the other hand, the beams appear to be moving at the same speed to the observer standing beside the track of the inter-galaxy train. However, since the train is moving, the light beam emitted in the direction in which the train is moving has to travel a longer distance than the one emitted in the opposite direction. Therefore, the beam emitted in the opposite direction reaches the target earlier than the beam emitted in the moving direction. Here, we have a contradiction.

If one contemplates on this paradox in a Humean fashion, this might not be a contradiction since according to Hume, "The ideas of space and time are therefore no separate or distinct ideas, but merely those of the manner or order, in which objects exist." Like the relative speeds of the balls appearing different to different observers, the timings at which light beams are emitted could be different for different observers. This is because an observation takes place when the light reaches the eyes of an observer. Einstein learned the relativity of observation and that of simultaneity from Hume and solved the paradox: when the observer inside the train observes the two light beams emitted simultaneously hit the target at the same time, the observer outside the train observes that "the light beam in the moving direction is emitted first, and the two beams hit the target at the same time." This was the moment when the science of matter received feedback from the science of man.

12.5 Remarks

We set forth a theory of man that tries to understand the world by constructing a model. We take experiences, or chunks of impressions, as primitives of the theory. A model is something that is constructed by agents. In doing so, agents use axioms such as coherence, according to which an agent can explain his own experiences, the behavior rule with respect to a solution concept, and "simplicity."

Inductive and deductive game theory should not be regarded as substitutes, rather, they can be viewed as complements. In reality, people use both induction and deduction in accumulating knowledge, which eventually affects their behavior. Furthermore, the construction of a model does not have to precede an action taking place. In fact, experiences include those impressions that are obtained through one's own behavior.

The case of eyeglasses presented in Sect. 5.1 may be understood better if we use an inductive inference rather than a deductive one. If a student does not understand what a teacher write on the blackboard, there are various possible explanations. In the rural areas during Meiji era, nobody had ever seen eyeglasses. It did not occur to teachers that a student who had no problem with eyesight in daily life may not see letters written on the board. Given such limited knowledge, an acceptable explanation was that the student was not smart enough to understand what the teacher asked her to do.

If we say a "model of the world," it may sound as though there existed an object called "the world." Although we do not know whether there exists a situation called an objective world or not, the concept of an objective game itself is a creation of researchers. An "objective game" is constructed by a researcher in order to understand our experiences/impressions better than otherwise. However, in the present framework, we do not have to presume the existence of such an objective world, nor do we have to take a position against it. Without entering into such a metaphysical discourse, the present framework can be used to address issues that the current society confronts.

Different individuals create different worlds. According to Encyclopaedia Britannica (2011), this idea can be seen in *Vijñānavāda* (the doctrine of consciousness), a school of *Mahāyāna* (greater vehicle), founded by Asanga and Vasubandhu (fifth century AD). They used the parable "*Issui-Shiken*," or "One water, four appearances": what humans view as "water" may be viewed as a "bloody sea" by *gaki* (hungry ghosts), as a "residence" by fish, and as a "land of treasure" by heavenly beings.

Our experiences are limited in various ways. We cannot feel what others feel. All we can do is to infer others' feelings from circumstances and their facial and other expressions. When we do these things, we have already constructed a model of others. In this sense, it may well be the case that animals other than humans have some ability to construct a model. After all, God "created man in his own image" (Genesis 1:27) as believed in the West, while humans and animals transmigrate into each other in the East.

Chapter 13
Segregation and Discrimination in Festival Games

13.1 Introduction

Societies consist of several groups.[1] Groups may be based on race, religion, culture, the degree of disability, and so forth. In these societies, the phenomena of discrimination and prejudices are typically observed. These phenomena raise not only practical societal issues but also offer some theoretical problems for economics and game theory. Among these problems is the treatment of interactions between behavioral and mental attitudes. The purpose of this chapter is to present a theoretical framework that enables us to analyze the relationships between these two components of multi-group societies. In this chapter, we look at the nature of discrimination and prejudices, and argue that it is not captured in the standard framework of economics and game theory. In spite of its wide applicability, let us use the term "ethnic" groups in the following argument just for the sake of convenience.

Discrimination is an overt attitude toward some ethnic groups. It is a certain mode of behavior that includes, as an example, denial of a minority's access to political power and economic opportunities. On the other hand, prejudices, which can be defined as associations of a certain group of people or objects with some negative traits, are covert in nature; they are beliefs or preferences as opposed to behavior. Unlike the beliefs and preferences typically assumed in economics, prejudices have some notable characteristics. They are categorical and generalized thoughts. They are usually caused by a lack of sufficient knowledge about the targeted people or objects. If we carefully listen to a negative opinion against a certain group of people, we often find that the person who expresses such an opinion has not met so many people of that group so as to make a logical claim. Generalization of limited knowledge to a categorical judgment is an important characteristic of prejudices. Another related characteristic of prejudices is that they contain fallacious elements to a significant degree.

---

[1]This and the next chapters are based on Kaneko and Matsui (1999).

© Springer Nature Singapore Pte Ltd. 2019
A. Matsui, *Economy and Disability*, Economy and Social Inclusion,
https://doi.org/10.1007/978-981-13-7623-8_13

In order to incorporate these characteristics in the scope of our research, we develop an analytical framework called inductive game theory. As its name suggests, induction is the key concept. In this theory, each player has little a priori knowledge of the structure of the society, but the lack of such knowledge is partially compensated for by the player's experiences in a recurrent situation. Here the player uses induction to derive an image of or a view of the society from these experiences. In this framework, we treat prejudices as a "fallacious" image against some ethnic groups. By focusing on the problem of discrimination and prejudices, we try to develop a theory of interactions between the thoughts in the mind of the player and his behavior in a social context. In the development, we do not discuss the information processing of the mind of the player; instead, we focus on logically possible images formed by induction in his mind.

An attempt to analyze fallacious beliefs and preferences in the existing frameworks of game theory poses some difficulty. To see this, we look at some existing theories, starting with the classical game theory of rational players and followed by learning and evolutionary theories.

In classical game theory since Nash (1951), it is often implicitly, and sometimes explicitly, assumed that players are rational in the sense of having high abilities of logical reasoning and knowledge of the structure of the game. Based on such abilities and a priori knowledge, the individual player makes a decision ex ante. We call this theory deductive game theory because deduction is the main process of reasoning. In this light, deductive game theory is appropriate for the study of societies where players are well informed for example, small games played by experts. Because the reasoning process of the rational player is always "correct" and is based on a priori knowledge, there is no room for the emergence of prejudices in deductive game theory. Moreover, the problem of prejudices could be addressed in this approach only if players are assumed to have false beliefs a priori.

Other theories in the literature are non-Bayesian learning and evolution. In the models of non-Bayesian learning, some prespecified learning rules are used to adjust players' beliefs and/or behavior. Players may learn some parameters of the game and strategies of others as well as the payoffs from their own behavior. In evolutionary game theory, the survival of the fittest is the main force in the selection of strategies. These approaches focus on economic problems where adaptive behavior and behavioral interactions are of prime importance. Inductive decision making is often their main focus, and little attention is paid to the formation of images or thoughts about the society in the minds of the players.

Figure 13.1 summarizes the major differences between these three types of theory. Arrows in the figure indicate the causality flows between knowledge (view) of structure and behavior.

Keeping the preceding discussions in mind, we describe our approach. We consider a specific game called the festival game, which is a variant of the game discussed in Kaneko and Kimura (1992). The festival game is a two-stage game in which the players are divided into several ethnic groups. These groups differ from each other only in their nominal ethnicities. In the first stage, all of the players simultaneously select a festival location. They observe which ethnic groups are present at their

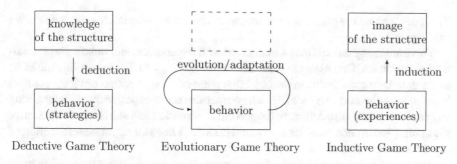

Fig. 13.1 Three approaches to game theory

respective festival locations, and then they simultaneously decide to take either a friendly or unfriendly action. We consider the situation where the festival game is played repeatedly.

In the repeated situation of the festival game, we consider a stationary state subject to occasional random trials. The probabilities of such trials are assumed to be small so that each player does not consider the events of simultaneous deviations of two or more players. In this environment, he accumulates his experiences from various unilateral deviations as well as from the stationary state. The experiences induced by the deviations of the individual player in question are called active experiences, and the ones induced by other players' unilateral deviations are called passive experiences.

In the absence of a priori knowledge, induction is taken as a general principle for the cognitive processes of the individual player. We consider two types of induction:

(i) inductive decision making, and
(ii) inductive construction of an individual image of the society.

The first is to choose a better strategy taught by experiences, and the second is to derive an interpretational view of the society based on experiences. The first type is categorized, more generally, into *inductive adjustments*, which could be found in classical equilibrium theory and learning theory in economics. The second type of induction is concerned not only with adjusting some parameters but also with building a new structure from experiences, which is the main focus of this chapter.

An individual player's image constructed inductively is formulated as a model of the society. Such a model is a partial description of the society including the individual's imaginary utility and observation functions. We give three coherence conditions on such a model, ones with the stationary, active, and passive experiences. These coherency conditions require that the utility and observation functions of the model generate the pieces of information corresponding to those experiences.

We consider another condition, called *rationalization*, for a model to satisfy. It follows from inductive decision making that the individual player makes a "rational" decision at every decision node that is reached. However, the action conceivable for him at such a decision node may lead to a social state never experienced. The rationalization condition requires that he rationalize his choice at such decision nodes.

This goes beyond the coherence requirement, since it is a restriction over states never experienced.

There are many models that are coherent with experiences and satisfy the rationalization condition. One example is the true-game model, which is essentially the same as the game we consider from the objective point of view. Another obvious example is the mere-enumeration model, which enumerates one's experiences without giving any causal relationship. Although the second is important as a start from the inductive viewpoint, we do not consider it explicitly, since it needs a slight generalization of our definition of a model.

The active experiences impose few restrictions on models other than utility maximization. Indeed, each deviation by an individual player induces only a single pair of a utility value and an observation of ethnicities. Ignoring the observed ethnicities, a player can always construct a simplistic model coherent with the active experiences in which the utility function depends only on his own actions. Such a model is called a naive hedonistic model. However, this model can rarely explain the passive experiences in a satisfactory manner.

When passive experiences are taken into account, the inductive construction of an individual view may involve prejudices. Since passive experiences are induced by other players' deviations, they are associated with the effects of the presence of other ethnic groups. A sophisticated hedonistic model uses the observed ethnicities as explanatory variables of one's utility. We show that this model explains the reality well in spite of its fallacy and exhibits preferential as well as perceptual prejudices.

In the literature, there are many works treating inductive reasonings in social contexts. To help the reader differentiate our inductive game theory from existing works, we mention three recent related theories: the case-based decision theory of Gilboa and Schmeidler (1995), the theory of self-confirming equilibrium of Fudenberg and Levine (1993), and the theory of subjective equilibrium of Kalai and Lehrer (1995). Then we mention an ancient but more directly related work: The allegory of the cave in Book VII of Plato's Republic (Plato 1941).

Case-based decision theory emphasizes the information processing of a decision maker who evaluates alternative choices based on similarities between the present problem and past cases under the assumption that similar experiences lead to similar effects. Such evaluations get adjusted as more cases become available. This is an individual decision theory based on inductive adjustments of similarity evaluations.

The self-confirming equilibrium of Fudenberg and Levine (1993) and the subjective equilibrium of Kalai and Lehrer (1995) describe a situation in which each player maximizes his expected payoff based on a belief consistent with what he observes in the course of play. Beliefs are expressed as subjective probabilities and are adjusted by new pieces of information obtained during the course of the game. These theories are also categorized into inductive adjustments, although they explicitly treat social aspects in contrast with the case-based decision theory.

The allegory of the cave in Book VII of Plato's Republic (Plato 1941) goes as follows. In the cave, prisoners have been, from childhood, chained by the leg and also by the neck, so that they cannot move and can see only the wall of the cave. On the wall, they see shadows of various things moving outside the cave, like the screen at a puppet-show. The only real things for them are the shadows. Each prisoner forms an individual view of the world from the shadows he has seen. Plato went on to discuss what might happen if one person were suddenly released to see the outside world, and how he would be treated after coming back to the other prisoners and telling them what he saw.

The framework of the present work, as well as its spirit, is similar to the story of the cave in that people with no a priori knowledge form a view of the society from experiences. The primary difference from the other works mentioned above is that the induction of our focus is one that builds a new structure out of limited experiences, while in the other works, parameters on the prespecified structures such as beliefs about other players' strategies are adjusted so as to be consistent with experiences.

We will divide the analysis into two chapters. This chapter is concerned with the deductive approach. Section 13.2 of this chapter considers a recurrent situation in which a festival game is played repeatedly. Section 13.3 states the basic postulates for our analysis. Section 13.4 characterizes the set of Nash equilibria of the festival game. Section 13.5 characterizes socially stable sets of the festival games.

13.2 Festival Games

We consider a recurrent situation where a game called the festival game Γ has been and will be played many times:

$$\text{unilateral trials}$$
$$\text{past} \ldots \Gamma \ldots \Gamma \ldots \Gamma \ldots \text{future}$$

We now provide a description of the festival game and some concepts to be used in the subsequent analysis.

The festival game Γ is a two-stage game. The player set $N = \{1, \ldots, n\}$ is partitioned into ethnic groups $N_1, \ldots, N_{e_0}$, where e_0 is the number of ethnic groups, and $|N_e|$ is the number of players in ethnic group N_e. Let $e(i)$ denote the ethnicity of player i, i.e., $i \in N_{e(i)}$. All the players are identical except for their ethnicities. There are ℓ locations for festivals. We may call the festival at location k $(k = 1, \ldots, \ell)$ *festival k*. Assume that $|N_e| > 2$ holds for all $e = 1, \ldots, e_0$.

The game Γ has two stages: the stage of choosing festival locations and the stage of acting in festivals (see Fig. 13.2). In the first stage of the game, each player simultaneously chooses a festival location. The choice of player i in this stage is denoted by $f_i \in \{1, \ldots, \ell\}$. We write $f = (f_1, \ldots, f_n)$.

After the choice of festival, each player observes the ethnicity configuration in the festival he chose, i.e., which ethnic groups are present in her festival. Formally, given $f = (f_1, \ldots, f_n)$, player i observes the ethnicity configuration of festival f_i, which is defined to be the set:

$$\mathcal{E}_i(f) = \{e(j) | f_j = f_i, \ j \neq i\}.$$

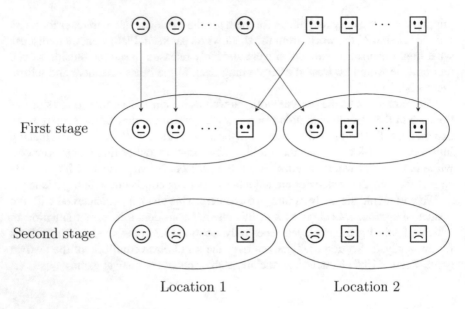

Fig. 13.2 Festival game with two groups

Each player can distinguish neither the identity of each participant nor the number of the participants of each ethnic group in the festival he chose. This is an assumption to simplify the subsequent analysis. Note that the ethnicity of player i is not counted if no other players of the same ethnicity are in the festival.

In the second stage, after observing the ethnicity configuration $\mathcal{E}_i(f)$ of festival f_i, player i chooses his attitude, either *friendly* or *unfriendly*, denoted by $(\smile)$ and $(\frown)$, respectively. Following the standard game theory, a choice in the second stage is expressed by a function

$$r_i : \{1, \ldots, \ell\} \times 2^{\{1, \ldots, e_0\}} \to \{(\smile), (\frown)\}.$$

A value $r_i(k, \mathcal{E})$ is player i's attitude in festival k if he observes the ethnic configuration $\mathcal{E}$.

A strategy of player i is a pair (f_i, r_i) of choices in the first and second stages. We write $r_i(f) = r_i(f_i, \mathcal{E}_i(f))$ and $r(f) = (r_1(f), \ldots, r_n(f))$. Let Σ_i be the set of strategy distributions of type i players. We write $\Sigma \equiv \Sigma_1 \times \cdots \times \Sigma_n$. For a pure strategy profile (f, r), the realization path is given by a pair $(f, r(f))$. We write $\sigma_i \in \Sigma_i$ and $\sigma \in \Sigma$.

Given a strategy profile (f, r), each player's payoff is determined by his attitude and the mood of the festival chosen by himself. The mood of festival f_i for player i is given by the number of friendly people in festival f_i other than player i himself, i.e.,

$$\mu_i(f, r) = \sum_{f_j = f_i, j \neq i} \mathbf{1}[r_j(f) = (\smile)], \tag{13.1}$$

where $\mathbf{1}[P]$ is an indicator function, i.e., $\mathbf{1}[P] = 1$ if P is true and 0 otherwise. We define the payoff function of player i as

$$U_i(f, r) = \mathbf{1}[r_j(f) = (\smile)](\mu_i(f, r) - m_0), \tag{13.2}$$

where m_0 is assumed to be a noninteger number greater than 2. Thus, the unfriendly action always induces the zero payoff, the payoff from the friendly action is increasing in the number of friendly people in the same location, and m_0 is the threshold beyond which the friendly action is preferred to the unfriendly action. Denote by $U_i(\sigma)$ the expected payoff of type i players when $\sigma \in \Sigma$ is taken.

A strategy profile $\sigma^* \in \Sigma$ is said to be a *Nash equilibrium* if for all $i \in N$ and all $\sigma_i \in \Sigma_i$,

$$U_i(\sigma^*) \geq U_i(\sigma^*_{-i}, \sigma_i),$$

where $(\sigma^*_{-i}, \sigma_i)$ denotes the strategy profile obtained from σ^* by replacing σ_i^* with σ_i. For the game Γ, we have the following equivalent definition of Nash equilibrium: for all $i \in N$,

$$U_i(\sigma^*) \geq U_i(\sigma^*_{-i}, (f_i, \delta_i)),$$

holds for all $(f_i, \delta_i) \in \{1, \ldots, \ell\} \times \{(\smile), (\frown)\}$, where δ_i can be identified with a constant strategy taking value δ_i in the second stage.

We have formulated the festival game Γ and relevant game theoretic concepts in the standard manner. However, because we do not follow the standard ex ante view, we should be careful about the interpretation of each concept. For example, the payoff function $U_i(\cdot)$ is not known to player i himself as a function; instead, only each value is perceived by him. Also, we should be careful about the use of the standard definition of strategy because, being a complete list of contingent actions, it appears to presuppose the knowledge of the extensive form of Γ. However, we can avoid this interpretation, and each player can "play" the game without being aware of the full-fledged concept of strategies.

13.3 Experiences and Inductive Stability

Section 13.3 describes the basic postulates for our analysis of the entire recurrent situation. Then we give the definitions of active and passive experiences for each individual player, and characterize the set of Nash equilibria.

In the recurrent situation of the game Γ, we consider a stationary state (strategy profile) σ^*, subject to unilateral deviations of individual players from σ^*. Unilateral deviations give some knowledge about the society's responses, and under certain postulates, such knowledge enables each player to "maximize" his payoff against the stationary state and leads to a Nash equilibrium.

We first describe the basic postulates behind our mathematical formulation. Some postulates are often presupposed in many game theoretical works. However, in order to emphasize what is different from these standard works and what is not, we make some of the underlying assumptions explicit and write them in the form of postulates. Postulate 1 (Knowledge structure):

(a) After each play of game Γ, player i observes only his utility value, $U_i(\sigma)$, if the game is played according to σ, in addition to the information she obtained during the play of the game.
(b) Player i knows that there are festival locations $1, \ldots, \ell$ for her first choice, and that the player has two options, friendly and unfriendly actions $((\smile), (\frown))$ in the festival she chose.

Other than this knowledge, the players are entirely ignorant of the structure of the game they play, including the player set N. In particular, although player i has the payoff function $U_i(\cdot)$, she does not know it as a function but receives a realized payoff value after each play of the game.

After each play of the festival game, player i has gained an *experience* that is a collection of the information he obtained through the course of the play. By Postulate 1, an experience is given by a quadruple:

$$[f_i, \delta_i, \mathcal{E}; h_i]$$

where f_i is the festival location that player i went to, δ_i is her own attitude, $\mathcal{E}$ is the ethnicity configuration she observed, and h_i is the payoff received.

We assume that players usually follow their behavior patterns σ^*, and that they make certain experiments and record the information obtained from such experiments. The following postulate makes this assumption explicit.

Postulate 2 (Behavior patterns and experiments):

(a) Given a stationary state σ^*, each player i behaves according to his behavior pattern σ_i^* subject to occasional trial deviations with small probabilities, but after each trial she returns to her own behavior pattern σ_i^*.
(b) Events of trials simultaneously made by two or more players have negligible frequencies, and they are ignored by the players.

When the experiences for player i show that it might be better to deviate, he would intentionally change his behavior pattern, which will be stated by another postulate.

In view of Postulate 2, the individual experiences in the past are categorized into the following three classes:

(s) stationary experience: the experience induced by the stationary state $\sigma^* = (f^*, r^*)$;
(a) active experiences: the experiences induced by one's own deviations;
(p) passive experiences: the experiences induced by deviations of some other players.

Each of the active and passive experiences is attained by the strategy profile induced by a unilateral deviation of a single player from σ^* by Postulate 2(b).

The focus of this analysis is not on the structure described in Postulate 2, but is on the step next to Postulate 2; that is, the focus of the present analysis is on possible individual views about the society constructed from experiences. Hence a mathematical treatment of these three types of experiences is crucial in the present analysis.[2]

In the following, a pure strategy distribution, a strategy distribution that puts all the weights on a pure strategy (f, r), is written as $\sigma = (f, r)$ and so forth. The *stationary experience* for player i under $\sigma^* = (f^*, r^*)$ is expressed as

$$(S): \quad s_i(\sigma^*) = [f_i^*, r_i^*(f^*), \mathcal{E}_i(f^*); U_i(\sigma^*)].$$

This is the collection of information that player i has regularly observed. Here f_i^* and $r_i^*(f^*)$ are the actions of player i, $\mathcal{E}_i(f^*)$ is the ethnicity configuration observed after the first stage, and $U_i(\sigma^*)$ is the payoff value received after the second stage. Note that player i is not aware of the expressions in the brackets, i.e., only the values described by these meta-expressions are observed by player i.

With a small frequency, player i deviates from her own behavior pattern σ^* and learns an active experience. An *active experience* under σ^* induced by a trial $(f_i, \delta_i) \in \{1, \ldots, \ell\} \times \{(\smallsmile), (\smallfrown)\}$ of player i is given as

$$(A): \quad [f_i, \delta_i, \mathcal{E}_i(f_{-i}^*, f_i); U_i(\sigma_{-i}^*, (f_i, \delta_i))].$$

The third element, $\mathcal{E}_i(f_{-i}^*, f_i)$, is the ethnicity configuration player i observes in the festival f_i, and the fourth element, $U_i(\sigma_{-i}^*, (f_i, \delta_i))$, is the utility value enjoyed by the player when she chooses attitude δ_i in festival f_i. Let $A_i(\sigma^*)$ denote the set of all active experiences of player i. Note that the stationary information $s_i(\sigma^*)$ is contained in $A_i(\sigma^*)$.

The passive experiences for player i are the experiences induced by a player other than player i herself. Formally, a *passive experience* under σ^* induced by a trial

---

[2]We do not fully specify the time structure and timing of trials. Although such a specification is not used in the present analysis, it would help in understanding the above argument to specify such possible time structures.

One possible formulation is to have a discrete time structure $\{\ldots, -2, -1, 0, 1, 2, \ldots\}$. Each player's behavior is subject to a stochastic disturbance, and if such a disturbance occurs then her behavior (f_i, δ_i) is randomly chosen. One possible assumption is that each disturbance occurs, with a small probability $\varepsilon > 0$ in each period, independently across the players. Then the probability of two or more players making simultaneous trials is at most of the second order. It means that the frequency of such trials is negligible relative to that of unilateral trials when ε is very small. Then player i collects the experiences of the first order.

Another model can be regarded as the limit of the above discrete time structure as the time interval tends to zero. The time structure is expressed as the real continuum $(-\infty, \infty)$. The festival game is played at each point in time. All players behave according to their stationary state σ^* at every point in $(-\infty, \infty)$, except for occasional disturbances, which make players try other actions. For each player, these disturbances follow a Poisson process. The Poisson processes are assumed to be independent across the players. Therefore, there is at most one trial made at each point in time with probability one, and negligible frequency of simultaneous trials is a consequence of this process.

$(f_j, \delta_j) \in \{1, \ldots, \ell\} \times \{(\smile), (\frown)\}$ of player $j \neq i$ is given by

$$(P): \quad [f_i^*, r_i^*(f_{-j}^*, f_j), \mathcal{E}_i(f_{-j}^*, f_j); U_i(\sigma_{-j}^*, (f_j, \delta_j))].$$

We denote the set of all passive experiences of player i by $P_i(\sigma^*)$.

Notice that there is the following asymmetry between the active and passive experiences. Player i notices that the differences between the stationary and active experiences were caused by his own deviations. However, by Postulate 1, he does not identify any other player to cause the difference between the stationary and passive experiences; he only sometimes receives different information.

A generic element of $A_i(\sigma^*) \cup P_i(\sigma^*)$ is denoted by $[\phi_i, h_i]$, where ϕ_i consists of f_i, δ_i and $\mathcal{E}$.

Each player i does not know her own utility function. However, player i has experienced various utility values in $A_i(\sigma^*) \cup P_i(\sigma^*)$. If player i has found a higher utility value than the stationary value, and if this value can be induced by the trial of player i, then she would have an incentive to increase the frequency of this deviation from the present stationary behavior σ_i^*. Therefore, we make a postulate on his behavior in such a case, which defines the stability of σ^*.

Postulate 3 (Inductive decision making):

(a) If no active experience in $A_i(\sigma^*)$ gives a higher payoff to player i than the stationary payoff $U_i(\sigma^*)$, then the player continues to play σ_i^* (still subject to occasional trials);

(b) If some active experience $[\phi_i; h_i] \in A_i(\sigma^*)$ gives a higher payoff to player i than the stationary payoff $U_i(\sigma^*)$, then the player would intentionally increase (either slightly or drastically) the frequency of the deviation inducing $[\phi_i; h_i]$.

The following definition is based on this postulate. We say that a player i has an *incentive for an intentional deviation* in σ^* if there is an active experience $[\phi_i; h_i] \in A_i(\sigma^*)$ that satisfies $h_i > U_i(\sigma^*)$. A strategy profile σ^* is *inductively stable* if no player has an incentive for an intentional deviation.

Proposition 13.1 *A strategy profile $\sigma^* \in \Sigma$ is inductively stable if and only if it is a Nash equilibrium in Γ.*

Proof By the definition of inductive stability, $\sigma^* \in \Sigma$ is inductively stable if and only if for any player i, $U_i(\sigma^*) \geq h_i$ holds for all $[\phi_i; h_i] \in A_i(\sigma^*)$. This statement is equivalent to $U_i(\sigma^*) \geq U_i(\sigma_{-i}^*, \sigma_i)$ for all $\sigma_i \in \Sigma_i$. $\qquad \square$

Inductive stability is simply a translation of the mathematical definition of Nash equilibrium. However, it is important to evaluate the claim of Proposition 13.1 from the viewpoint of inductive decision making.

The *if* part means that if player i has no experience with a utility value higher than that in the stationary state, then player i continues playing his strategy, which is Postulate 3(a). Hence if no player has actively experienced a higher utility value, then σ^* is stable in the sense that all the players continue playing σ^*. This part involves a weak form of induction: when the player has experienced the same stationary information except for some occasional changes, he expects that if he does not change his action, nothing else will change, either.

The *only-if* part is more substantive. If a player has an active experience with a higher utility value than that in the stationary state, then the player intentionally changes his behavior, either slightly or drastically—Postulate 3(b). In this sense, σ^* is no longer stationary. Here the player does not fully understand the possible consequences of his intentional deviations. The player is making an inductive decision based on a generalization of his active experiences, and he expects to receive a higher utility more frequently by making that deviation more often than before.

Finally, we also comment on sequential rationality—the equilibrium requirement for the second stage.[3] The essential part of sequential rationality in the festival game Γ is the equilibrium requirement for the reactions of the players to a deviation by a single player; for example, the reactions of the players in a festival must be rational when an outsider attends the festival. The payoff maximization for such a reaction requires an insider to have trial responses to the deviation of the outsider. Thus, sequential rationality needs experiences induced by trials of two or more players. However, Postulate 2(b) assumes that those events are negligible for each player. Thus, we cannot assume sequential rationality in our context. Section 14.6.1 will consider the problem of sequential rationality again.

13.4 Segregation and Discrimination

Before going to the main part of inductive game theory, we consider the structure of Nash equilibria in the festival game Γ. There are three types of equilibria, one of which exhibits segregation of some ethnic groups and discriminatory behavior to support such segregation. The other two are degenerated equilibria. The following theorem characterizes the set of Nash equilibria, *a fortiori*, inductively stable profiles.

Theorem 13.2 *A strategy profile* $\sigma^* = (\sigma_1^*, \ldots, \sigma_n^*) = ((f_1^*, r_1^*), \ldots, (f_n^*, r_n^*))$ *is a Nash equilibrium if and only if for any* $i \in N_e$ *and* $e = 1, \ldots, e_0$:

(a) *if* $\mu_i(\sigma^*) \geq m_0$, *then* $f_j^* = f_i^*$ *holds for any* j *with* $e(j) = e$, *and* $r_k^*(f^*) = (\smile)$ *holds for any* k *with* $f_k^* = f_i^*$;
(b) *if* $\mu_i(\sigma^*) \geq m_0$, *then* $\mu_i(\sigma^*) \geq \mu_i(\sigma_{-i}^*, (f_i, (\smile)))$ *holds for any* $f_i = 1, \ldots, \ell$;
(c) *if* $\mu_i(\sigma^*) < m_0$, *then* $\mu_i(\sigma^*) = 0$ *holds, i.e.,* $r_k^*(f^*) = (\frown)$ *holds for any* k *with* $f_k^* = f_i^*$;
(d) *if* $\mu_i(\sigma^*) < m_0$, *then* $\mu_i(\sigma_{-i}^*, (f_i, (\smile))) < m_0$ *holds for any* $f_i = 1, \ldots, \ell$.

Suppose that the number of friendly people at f_i^* exceeds the threshold m_0. Claim (a) states that every player of the same ethnicity as player i goes to the same festival, and every player in this festival takes the friendly action. Claim (b) states that if player goes to a different location f_i, then the number of friendly people at

---

[3] See Fudenberg and Tirole (1991) and Mas-Colell et al. (1995) for more detail.

f_i becomes smaller than or equal to the number at f_i^*. Note that (a) allows more than one ethnic groups to go to the same festival (see Figs. 13.3 and 13.6). On the other hand, suppose the number of friendly people at f_i^* is less than the threshold m_0. Then, Claim (c) states that no player at f_i^* takes a friendly action (see festivals 2 and 3 in Fig. 13.6), and Claim (d) states that wherever player i may go, the number of friendly people would not exceed the threshold m_0.

Proof (only-if part) Suppose that a strategy profile $\sigma^* = (\sigma_i^*)_{i \in N} = (f_i^*, r_i^*)_{i \in N}$ is a Nash equilibrium. We prove (a)–(d) one by one. To prove (a), suppose that the number of friendly people at f_i^* exceeds the threshold m_0. Then player i obtains a positive payoff by taking the friendly action. Suppose that player j of the same ethnicity as player i goes to $f_j \neq f_i^*$. If player j joins f_i^*, the ethnicity configuration at f_i^* does not change except possibly for i, and therefore, player j would not trigger any different action in the second stage except for the action of player i and obtains no less than $U_i(\sigma^*)$, which implies $U_j(\sigma^*) \geq U_i(\sigma^*)$. Using the same logic, all the players of ethnicity e obtains $U_i(\sigma^*)$ in equilibrium. Due to our assumption, there exists at least one location, ℓ', such that at least two players of ethnicity e choose ℓ' under σ^* (ℓ' may be f_i^* or another location). Then there exists a player, say k, of ethnicity e who chooses ℓ'' under σ^*. If player k goes to ℓ', then the ethnicity configuration at ℓ' does not change for everyone there. Thus, k obtains $U_i(\sigma^*) + 1$ by such a deviation, which is a contradiction. Also, if player i obtains a positive payoff at f_i^*, everyone there should be able to obtain a positive payoff by taking the friendly action, while the payoff is zero if one takes the unfriendly action. Thus, everyone takes the friendly action. Claim (b) is a direct consequence of Nash equilibrium.

To prove (c), suppose now that the number of friendly people at f_i^* is less than the threshold m_0. Then anyone taking the friendly action would obtain a negative payoff. Thus, the number of the players taking the friendly action is zero at f_i^*. Claim (d) is again a direct consequence of Nash equilibrium.

(if part) Suppose that (a)–(d) hold. Take player i. Consider two cases. First, suppose that the number of friendly players exceeds m_0. Then (a) implies that player i takes the friendly action and obtains a positive payoff. Moreover, player i cannot gain by deviation due to (b) and the fact that taking the unfriendly action leads to zero payoff. Thus, player i has no incentive to deviate in this case.

Second, suppose that the number of friendly players at f_i^* is less than m_0. Then (b) implies that player i takes the unfriendly action. Condition (d) implies that no matter where player i may go, the player cannot obtain a positive payoff. Thus, player i has no incentive to deviate.

Given that the above statement is true for an arbitrary player, the strategy profile is a Nash equilibrium.   □

The above theorem enables us to classify the set of equilibria into the following three classes:

Integration equilibria: $f_i^* = f_j^*$ and $r_i^*(f^*) = (\smile)$ hold for all $i, j \in N$: all the players choose the same location and take the friendly action. The players enjoy the best payoff (see Fig. 13.3). In Fig. 13.3, the height of the column corresponds to the

Fig. 13.3 Integration
equilibrium

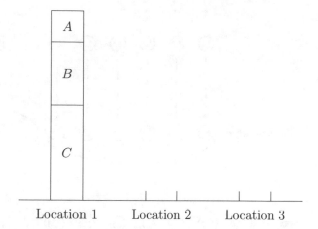

number of the players of the ethnic groups. In this case, there are three ethnic groups, A, B, and C. Among them, C is the largest, B the second largest, and A is the smallest group.

Segregation equilibria: $f_i^* \neq f_j^*$ and $\mu_i(\sigma^*) \geq m_0$ for some $i, j \in N$: some players of different ethnicities go to different festivals and at least one festival is active. Segregation occurs in this equilibrium (see Figs. 13.6 and 13.7).

No active festival equilibria: This is a trivial equilibrium, wherein no player takes a friendly action on the equilibrium path.

13.4.1 An Example

Let us illustrate how segregation is sustained in equilibrium. For the sake of illustration, consider a society with 10 players, i.e., $N = \{1, \ldots, 10\}$, with $N_1 = \{1, \ldots, 6\}$ being a majority group and $N_2 = \{7, \ldots, 10\}$ being a minority group. Let $m_0 = 2.5$. Consider a segregation equilibrium where in the stationary state, all the players in N_1 go to location 1, while all the players in N_2 go to location 2. Suppose further that they all take the friendly actions in the stationary state. Figure 13.4 shows this situation. In this outcome, each player in group 1 obtains 2.5 ($= 5 - m_0$). On the other hand, each player in group 2 obtains 0.5 ($= 3 - m_0$).

There are a variety of deviation types. However, the major, and perhaps unique, concern among such deviations is that a player in group 2 might migrate into location 1 instead of staying in location 2. Figure 13.5 illustrates how the equilibrium wipes off such an incentive. It shows that if a player in group 2 goes to location 1, then some, if not all, players in group 1 observe a change in ethnic configuration and alter their second-stage behavior from the friendly action to the unfriendly action. The figure shows that four of them make such changes. Because of this change, the player from group 2 cannot obtain a higher payoff than the payoff in the stationary

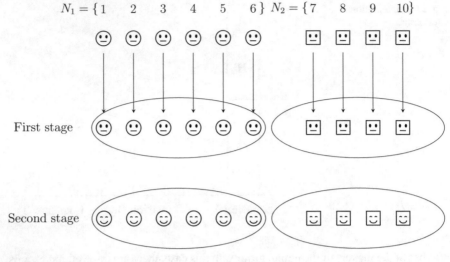

Fig. 13.4 Segregation equilibrium: the stationary state

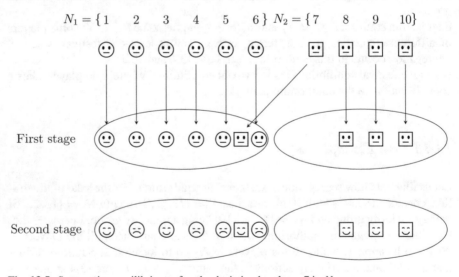

Fig. 13.5 Segregation equilibrium: after the deviation by player 7 in N_2

state even if this player takes the friendly action. This way, the stationary state with segregation sustains itself as an equilibrium outcome.

If there are three ethnic groups, we may have a variety of patterns of segregation (see Figs. 13.6 and 13.7 as examples). In Fig. 13.6, the players in both A and B go to location 1 and enjoy the positive payoff, while the players in C go to locations other than location 1 and take the unfriendly action. In order for this outcome to be an equilibrium outcome, it must be the case, among others, that no player in C has

Fig. 13.6 Partially active and partial segregation equilibrium

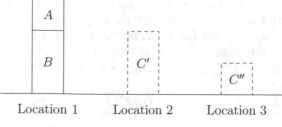

Fig. 13.7 Fully active and complete segregation equilibrium

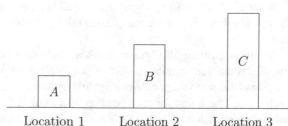

an incentive to choose location 1 where the players in two other ethnic groups enjoy a positive payoff. If a player in C goes to location 1, it would change the ethnicity configuration. In this off-path, the number of the players taking the friendly action does not exceed the threshold m_0. If this is the case, then no player in C has an incentive to go to location 1. Segregation is maintained by discriminating the players of unwelcome ethnicity.

In Fig. 13.7, there are several active festivals, some of which are larger than others. A player who goes to a small festival would have an incentive to visit a larger festival if the players at the larger festival took the friendly action in response to the player's participation. However, if the player goes to a larger festival and many players there take the unfriendly action, the payoff to the newcomer as well as to others in the larger festival decreases considerably. Like in the case of Fig. 13.6, the equilibrium is maintained by discrimination.

The following corollary states that in a segregation equilibrium, some players at festival f_i^* respond to the participation of player j in an unfriendly manner if j is from a smaller festival.

Corollary 13.3 *Let $\sigma^* = (f^*, r^*)$ be a segregation equilibrium. Suppose $f_i^* \neq f_j^*$ with $\mu_i(\sigma^*) > \mu_j(\sigma^*)$, and let $f_j = f_i^*$.*

(i) If $\mu_j(\sigma^*) \geq m_0$, then $\mu_j(\sigma_{-j}^*, (f_j, (\smile))) \leq \mu_j(\sigma^*)$.
(ii) If $\mu_j(\sigma^*) = 0$, then $\mu_j(\sigma_{-j}^*, (f_j, (\smile))) < m_0$.

In (i), if player j of an active festival smaller than f_i^* comes to f_i^*, the number of the induced friendly actions by her presence is not more than that of f_j^* to which player j regularly goes. Thus, discrimination necessarily occurs when a player of a smaller festival comes to a larger festival. The difference $\mu_j(\sigma^*) - \mu_j(\sigma_{-j}^*, (f_j, (\smile)))$ is the number of players switching from friendly to unfriendly actions in response to the presence of player j at festival f_i^*. Discrimination may or may not occur when a player in a larger festival visits a smaller one.

In (ii), festival at f_j^* is inactive, and then the induced mood must be worse than the threshold m_0. To simplify the subsequent argument, we focus on the fully active equilibria $\sigma^* = (f^*, r^*)$ that satisfy:

$$\textbf{FA: } r_i^*(f^*) = (\smile) \text{ for all } i \in N.$$

Condition FA allows some segregation equilibria, but eliminates some others such as the one in Fig. 13.6, where some players take the unfriendly action on the equilibrium path.

When a Nash equilibrium σ^* satisfies subgame perfection, discriminators and non-discriminators cannot coexist in one festival. In particular, all the players in a large festival have to discriminate against those who come from a smaller festival in such a case. Note that for each Nash equilibrium, there is an equilibrium satisfying subgame perfection such that their realization paths are identical.

13.5 Escaping from Segregation

This section uses socially stable sets discussed in Chap. 8 to solve for socially stable sets in the festival games.[4] As the title of the section indicates, it is shown that segregation equilibrium is not in any socially stable set. To be precise, in order to justify the best response dynamics in a rigorous manner, we need to construct a random matching model. Here, we remain vague about the setup as to how they repeat the situation.

Recall that each strategy profile is written as (f, r). Then let us give a modified version of the best response path defined in Sect. 8.3.4. Given $\sigma_0 \in \Sigma$, a best response dynamic path starting from σ_0 is a continuous function p from some time interval $[0, T]$ to the set of strategy distributions that moves toward the best response strategy b at each point in time. The formal definition is given below.

Definition 13.1 Given a strategy profiles $\sigma_0 \in \Sigma$ and $T \in (0, \infty)$, a continuous function $p : [0, T] \to \Sigma$ is a best response dynamic path starting from σ_0 if p is differentiable from the right, $p(0) = \sigma_0$, and there exists a step function $b : [0, 1) \to \Sigma$ continuous from the right such that

$$\frac{d^+}{dt} p_k(t) = \alpha_k [b_k(t) - p_k(t)], \quad \alpha_k \geq 0,$$

and

$$b_k(t) \in BR_k(p(t))$$

---

[4]This section is not in Kaneko and Matsui (1999).

hold for all $k \in N$ and all $t \in [0, T)$.

This definition implies that the behavior pattern may go only in the direction of the present best response. The difference between the present definition and the one in Sect. 8.3.4 is the speed at which each ethnic group responds to the current state, i.e., α_k values may be different for different k values in the present definition, while α_k values are equal to each other in the definition of Sect. 8.3.4.

Next, we have the following definition of accessibility. First, using best response paths, we say that a strategy distribution g is *directly accessible* from f if there exists a best response path $p : [0, T] \to \Sigma$ for some $T > 0$ such that $p(0) = f$ and $p(T) = g$. The notion of accessibility is then recursively defined. We say that g is accessible from f if at least one of the following is satisfied: (i) g is directly accessible from f; (ii) there exists a sequence (g_n) converging to g such that g_n is accessible from f for all $n = 1, 2, \ldots$; and (iii) g is accessible from h which in turn is accessible from f. Now, our solution concept is as follows.

Definition 13.2 A nonempty subset F^* of Σ is called a socially stable set (with respect to the best response dynamics), or SS set(BR), if

(i) any $g \notin F^*$ is not accessible from any f in F^*; and
(ii) every $g \in F^*$ is accessible from every f in F^*.

A socially stable set is stable in the sense that once the actual behavior pattern falls in the set, another strategy distribution may be realized if and only if it is contained in the set.

Given a location $\ell' = 1, \ldots, \ell$, let $\Sigma^{\ell'}$ be the set of strategy profiles according to which every player goes to location ℓ' and take the friendly action, i.e.,

$$\Sigma^{\ell'} = \{ \sigma \in \Sigma | \forall (f, r) \in \sigma \, [f = (\ell', \ldots, \ell') \wedge \forall i \in N \, r_i(\mathcal{E}_i(f)) = (\smile)] \}.$$

The following theorem states that these sets, $\Sigma^{\ell'}$ sets, are the only socially stable sets.

Theorem 13.4 $\Sigma^{\ell'}$ *sets* $(\ell' = 1, \ldots, \ell)$ *are the only socially stable sets.*

The proof of the theorem is similar to the proof of optimality in games with cheap-talk. The idea is to use an empty location to induce cooperation.

Proof Take an arbitrary strategy distribution $\sigma^0 \in \Sigma$. To begin with, we construct a pure strategy distribution $[f, r] = ([f_i, r_i])_{i \in N}$ that is accessible from σ^0 in n steps. First, let (f_1, r_1) be the best response to σ^0. The more type 1 players take (f_1, r_1), the more attractive the strategy becomes relative to other strategies. Thus, $\sigma^1 = (\sigma^0_{-1}, [f_1, r_1])$ is accessible from σ^0. Recursively, at step $m = 2, \ldots, n$, $\sigma^m = (\sigma^{m-1}_{-m}, [f_m, r_m])$ is accessible from σ^{m-1}. By construction and transitivity, $\sigma^n = [f, r] = ([f_i, r_i])_{i \in N}$ is accessible from σ^0.

Let us classify $[f, r]$ into three families:

$$\Sigma^* = \{(f, r) \mid \forall i \in N \ U_i(f, r) = \max_{(f', r')} U_i(f', r')\};$$

$$\Sigma^0 = \{(f, r) \mid \forall i \in N \ \forall \ell' \ E[U_i(\sigma) \mid f_i = \ell'] \leq 0\};$$

$$\Sigma^1 = \Sigma \setminus (\Sigma^* \cup \Sigma^0).$$

In the above expression, $E[U_i(\sigma) \mid f_i = \ell']$ is the conditional expected payoff of player i taking strategy σ_i given the player i choice of ℓ'. Notice also that if a strategy profile is in Σ^1, then some players obtain positive conditional expected payoffs in some location, and therefore, the number of friendly players, i.e., those who take the friendly actions, exceeds the threshold m_0. In the following proof, all the strategy distributions are pure.

$[\sigma^n \in \Sigma^0]$ Suppose that σ^n is in Σ^0. Then every player's best response is to take the unfriendly action at any reached information set. Repeat the same procedure as before to obtain $\bar{\sigma}$ in which no player takes the friendly action at any reached information set, and it is accessible from σ^n.

Next, pick an ethnic group, say, N_1. Consider a move from $\bar{\sigma}$ toward $\bar{\sigma}^0$ where the players in N_1 choose location 1, while others choose location 2 with the unfriendly action being taken at the reached information sets. Then $\bar{\sigma}^0$ is accessible from $\bar{\sigma}$.

Then, consider a state $\hat{\sigma}^0$ where the players in N_1 take a strategy such that they choose location 1, take the unfriendly actions as before if they observe N_1 only, but take the friendly actions if they observe people from some different ethnic groups. Then Since $\hat{\sigma}^0$ gives the exactly the same payoff as $\bar{\sigma}^0$, there exists a best response path that moves toward $\hat{\sigma}^0$ from $\bar{\sigma}^0$. A path $(1 - t)\bar{\sigma}^0 + t\hat{\sigma}^0$ is a best response path at least for $t \in [0, 1)$.

Now, since the players in N_1 are taking an accommodating strategy, every other player has an incentive to move to location 1 and take the friendly action. Indeed, by taking such a strategy, the mover would obtain at least $(|N_1| - m_0) > 0$. We call such a strategy profile $\hat{\sigma}$. Then $\hat{\sigma}$ is accessible from $\hat{\sigma}^0$. Thus, $\hat{\sigma}$ is accessible from σ^0 by way of transitivity.

$[\sigma^n \in \Sigma^1]$ Next, consider σ^n in Σ^1. In this case, there exist an ethnic group N_k and a location ℓ' such that the expected payoff of the players in N_k is the highest (there may be a tie) among all the pairs of $(N_{k'}, \ell'')$. Assume without loss of generality that $k' = \ell'' = 1$ holds. It must be the case that according to (f_1^n, r_1^n), the players in N_1 go to location 1 and take the friendly actions on the outcome path. Let (f_1^{n+1}, r_1^{n+1}), the players in N_1 go to location 1 and take the friendly actions both on and off the outcome path. This is a best response to σ^n since the change was made only off the outcome path. Then the strategy distribution $\sigma^{n+1} = (\sigma_{-1}^n, [f_1^{n+1}, r_1^{n+1}])$ is accessible from σ^n.

Then we essentially repeat the last paragraph of the previous case. Since the players in N_1 are taking an accommodating strategy, every other player has an incentive to move to location 1 and take the friendly action. Indeed, by taking such a strategy, the mover would obtain at least $(|N_1| - m_0) > 0$. We call such a strategy profile $\hat{\sigma}$. Then $\hat{\sigma}$ is accessible from σ^{n+1}. Thus, $\hat{\sigma}$ is accessible from σ^0 by way of transitivity.

Note that the destination is $\hat{\sigma}$ for both cases, and that $\hat{\sigma}$ is in Σ^* where everyone obtains the highest payoff. The remaining task is to show that from any strategy profile in $\Sigma^{\ell'}$ ($\ell' = 1, \ldots, \ell$), there is no path that leaves it. This is easily verified since if everyone gathers at ℓ' and takes the friendly action, then any player who either goes to a different location or takes the unfriendly action on the outcome path will decrease her own payoff. $\qquad\square$

The theorem itself makes a strong statement that the only socially stable set is of the form in which everyone eventually goes to the same location and takes the friendly action. In order to appraise this theorem, however, we need to examine its proof. The proof is a little complicated since it has to show that the efficient outcome is accessible from *any* strategy distribution. The logic of the proof is better understood if we focus on a path from a particular distribution.

Consider Example 13.4.1 again. We would like to show an accessible path from the segregation equilibrium examined therein to the integration equilibrium according to which every player goes to location 1 and takes the friendly action. Remember that in the segregation equilibrium, many, if not all, players of type 1 take the unfriendly action if a player of type 2 visits location 1. As long as the players of type 1 keep discriminating the type 2 players, they cannot visit location 1.

We need two steps to escape from the segregation equilibrium. The first step is to change the off-path behavior of the type 1 players. Since the reaction to the visit of type 2 players is an off-path event, any reaction can be a part of a best response. In this case, the type 1 players switch their behavior from friendly to unfriendly when they observe type 2 without changing other parts of the strategy profile. This strategy distribution is accessible from the segregation equilibrium.

The second step is from this strategy distribution to the integration equilibrium. This part is easy. Since the type 1 players take the friendly action against the type 2 players, they have an incentive to migrate into location 1 and take the friendly action. The best response of type 2 players is to take such a strategy. This way, the strategy distribution reaches the integration equilibrium.

This transition is accomplished only when the type 1 players stop discriminating against the others. Accommodating behavior of the majority is an essential part of a move toward integration.

Chapter 14
Prejudices Induced by Segregation

14.1 Introduction

This chapter is concerned with the inductive approach where the players of the game construct models to understand the game (society) they play. This chapter is a continuation of Chap. 13, and therefore, we continue the analysis directly. Section 14.2 provides the definitions of individual models and coherence with experiences. Section 14.3 examines rationalization that requires that the model has to justify the action taken in a stationary state. Section 14.4 considers naive and sophisticated hedonistic models. Section 14.5 considers the possibility of escaping from prejudicial models. Section 14.6 discusses implications of our analysis for the game theory, economics, and sociological literature, considers interactions between individual views and behavior, and finally discusses the traditional view of a game with common knowledge of the structure of the game.

14.2 Inductive Construction of an Image of the Society

In an inductively stable state $\sigma^* = (f^*, r^*)$, each player i has accumulated experiences $A_i(\sigma^*) \cup P_i(\sigma^*)$ via occasional trials. The player does not know the structure of the game Γ by Postulate 1, but may infer, from the player's experiences $A_i(\sigma^*) \cup P_i(\sigma^*)$, what has occurred in the society. Here we consider possible views about the society formed by player i from her experiences $A_i(\sigma^*) \cup P_i(\sigma^*)$. Here, we apply again an inductive principle to this process, which is considerably stronger than the induction used in Sect. 13.2. The player generalizes her experiences into an explanatory causal relationship and builds a model of the society. In this section, we provide a general definition of such a model and two requirements: one is coherence with experiences, and the other is rationalization.

© Springer Nature Singapore Pte Ltd. 2019
A. Matsui, *Economy and Disability*, Economy and Social Inclusion,
https://doi.org/10.1007/978-981-13-7623-8_14

14.2.1 Individual Models Built by a Player

An individual model, $\mathcal{M}$, of player i is given by a sextuple $\langle \hat{N}, \hat{Z}, \hat{o}_i, \hat{u}_i; x^0, X \rangle$, where

1. $\hat{N}$ is a finite set of *players*;
2. $\hat{Z}$ is a set of *potential social states*;
3. $\hat{o}_i$ is the *observation function*;
4. $\hat{u}_i$ is the *utility function*;
5. $x^0 \in \hat{Z}$ is the stationary social state;
6. $X \subset \hat{Z}$ is the set of realized social states with $x^0 \in X$.

These constituents are all imaginary in the sense that they are constructed in the mind of player i. The first four constituents, $\hat{N}, \hat{Z}, \hat{o}_i$, and $\hat{u}_i$, are intended to describe the basic structure of the society or game that player i imagines. They constitute an alternative description of a game in extensive form Γ except that the other players' payoffs are omitted.

The last two constituents, x^0 and X, correspond to the realization path of the stationary state σ^* and the terminal nodes induced by individual deviations. That is, x^0 and X describe the play of the game; in particular, X is the set of relevant states reachable from x^0 by unilateral deviations of player i and some other players.

The game Γ and an individual model $\mathcal{M}$ have significant differences in their cognitive bases. The former is the objective description of the society, and the latter is its subjective description in the mind of player i. In particular, we emphasize the following difference. In Γ, player i has the utility function $U_i(\cdot)$, which means that the player receives each realized utility value, but the player does not knows U_i as a function. On the other hand, because the player builds $\mathcal{M}$ in her mind, she perceives $\hat{u}_i$ as a function.

In the sequel, we make the following assumptions on $\mathcal{M} = \langle \hat{N}, \hat{Z}, \hat{o}_i, \hat{u}_i; x^0, X \rangle$.

Assumption M1: $\hat{N}$ is a set expressed as the union of disjoint sets, $\hat{N}_1, \ldots, \hat{N}_{e_0}$ with $i \in \hat{N}_{e(i)}$.

Assumption M2: $\hat{Z} = \{1, \ldots, \ell\}^{\hat{N}} \times \{(\smile), (\frown)\}^{\hat{N}} \times Y$, where Y is some arbitrary set.

Assumption M3: $\hat{o}_i(g, \delta, y) = (g_i, \delta_i, \hat{E}_i(g))$ for all $(g, \delta, y) \in \hat{Z}$, where $\hat{E}_i(g) = \{e(j) | g_j = g_i, j \neq i, j \in \hat{N}\}$ for any g.

Since $x^0 \in \hat{Z}$, x^0 can be expressed as (g^0, δ^0, y^0). This notation will be used throughout the section.

Assumption M1 means that $\hat{N}$ is the set of imaginary players partitioned into the ethnic groups $\hat{N}_1, \ldots, \hat{N}_{e_0}$, and player i belongs to the group $\hat{N}_{e(i)}$. Assumption M2 expresses the idea that player i knows that every player in $\hat{N}$ has the same action space as that of player i. The additional Y is the set of hidden parameters player i imagines. When Y is a singleton, M2 is essentially equivalent to $\hat{Z} = \{1, \ldots, \ell\}^{\hat{N}} \times \{(\smile), (\frown)\}^{\hat{N}}$.

Assumption M3 means that player i believes that she observes her own choice (g_i, δ_i) and the ethnicity configuration $\hat{E}_i(g)$. It is important to notice that M3 excludes the possibility that the observation function $\hat{o}_i$ depends on an additional variable $y \in Y$ (see Remark 1 below). On the other hand, we impose no further restriction on the utility function $\hat{u}_i$. Hence, $\hat{u}_i$ may depend on the additional variable y.

The additional space Y is the domain of an exogenous explanatory variable. The introduction of this space gives some freedom to the possible models. Nevertheless, since a model with a large domain Y gives up a fine causal explanation, the less dependent a model is on y, the stronger is its explanatory power. In the models considered in the sequel, this additional variable y is used to explain some utility changes.

To illustrate the above definition of an individual model, we give one example of a model called the *true game model* $\mathcal{TG}$. It is essentially a redescription of the festival game Γ itself in terms of the above language. It will be shown in Sect. 14.6.1 that this model hardly satisfies the second requirement of rationalization.

Let $\sigma^* = (f^*, r^*)$ be a strategy profile. The true-game model of player i is given as $\mathcal{TG} = \langle \hat{N}, \hat{Z}, \hat{o}_i, \hat{u}_i; x^0, X \rangle$:

TG1 $\hat{N}_e = N_e$ for all $e = 1, \ldots, e_0$;
TG2 $\hat{Z} = \{1, \ldots, \ell\}^{\hat{N}} \times \{(\smile), (\frown)\}^{\hat{N}}$;
TG3 $\hat{o}_i(f, \delta) = (f_i, \delta_i, \mathcal{E}_i(f))$ for all $(f, \delta) \in \hat{Z}$;
TG4 $\hat{u}_i(f, \delta) = \delta_i(\mu_i(f, \delta) - m_0)$ for all $(f, \delta) \in \hat{Z}$;
TG5 $x^0 = (f^*, r^*(f^*))$;
TG6 $X = \{x^0\} \cup X_A \cup X_P$, where

$$X_A = \bigcup_{(f_i, \delta_i)} \{((f^*_{-i}, f_i), (r^*_{-i}(f^*_{-i}, f_i), \delta_i))\},$$

$$X_P = \bigcup_{j \neq i, (f_j, \delta_j)} \{((f^*_{-j}, f_j), (r^*_{-j}(f^*_{-j}, f_j), \delta_j))\}.$$

The true-game model, $\mathcal{TG}$, is described by focusing on the terminal nodes in the game Γ. The observation function $\hat{o}_i$ gives the pieces of information obtained in the course of the play of Γ, and the utility function $\hat{u}_i$ gives the value equal to the payoff $\delta_i(\mu_i(f, \delta) - m_0)$ assigned to the corresponding terminal node in Γ. The stationary state x^0 corresponds to the realization path $(f^*, r^*(f^*))$ of $\sigma^* = (f^*, r^*)$. The set X of relevant social states contains the three types of states: a state in X_A is induced by a deviation of player i, a state in X_P is induced by a deviation of some other player. Thus, the first four constituents $\langle \hat{N}, \hat{Z}, \hat{o}_i, \hat{u}_i \rangle$ are an alternative description of the extensive form game Γ except for the absence of the other players' observation and utility functions. The last pair (x^0, X) corresponds to the stationary state σ^* and the terminal nodes induced by individual deviations. Consequently, the model $\mathcal{TG}$ satisfies Assumptions M1-M3. Note that the additional space Y is not used.

Remark 1 From the viewpoint of induction, there is another extreme and important example, which is an enumeration of the experiences without adding any additional structure. To have this example in our theory, we need to generalize Assumption M3 so that the observation function $\hat{o}_i$ depends on the additional variable y. Then all experiences, except the player's own actions, can be recorded in the space Y. We call such a model the mere-enumeration model $\mathcal{ME}$. We will refer to this example in the discussions in Sect. 14.6.2.

14.2.2   Coherence of Models with Experiences

This subsection formulates some requirements for a model to be coherent with the stationary, active, and passive experiences. For an understanding of these coherence requirements and of their uses, we give two theorems immediately after the formulation of these requirements, and we illustrate them by using the true-game model.

Let $\sigma^* = (f^*, r^*)$ be a stationary state, and $\mathcal{M} = \langle \hat{N}, \hat{Z}, \hat{o}_i, \hat{u}_i; x^0, X \rangle$ be an individual model of player i satisfying M1-M3. Stationary state $\sigma^* = (f^*, r^*)$ defines the stationary experience $s_i(\sigma^*)$, active experiences $A_i(\sigma^*)$, and passive experiences $P_i(\sigma^*)$. The coherence requirement states that model $\mathcal{M} = \langle \hat{N}, \hat{Z}, \hat{o}_i, \hat{u}_i; x^0, X \rangle$ generates these experiences.

Formally, the coherence condition for a model $\mathcal{M}$ with the stationary experiences is given as follows:

Condition CS: $[\hat{o}_i(x^0); \hat{u}_i(x^0)] = s_i(\sigma^*)$.

Recall that $\hat{o}_i = (g_i^0, \delta_i^0, \hat{E}_i^0(g^0))$ and $s_i(\sigma^*) = [f_i^*, r_i^*(f^*), \mathcal{E}_i(f^*); U_i(\sigma^*)]$. This requires that the stationary state x^0 in $\mathcal{M}$ gives the stationary information $s_i(\sigma^*)$ that player i has obtained under σ^*.

Second, the coherence condition for $\mathcal{M}$ with $A_i(\sigma^*)$ is formulated as follows:

Condition CA: For any $[\phi_i; h_i]$, $[\phi_i; h_i]$ is in $A_i(\sigma^*)$ if and only if there is a state $x = (g, \delta, y) \in X$ such that $g_{-i} = g_{-i}^0$ and $[\hat{o}_i(x); \hat{u}_i(x)] = [\phi_i; h_i]$ hold.

This states that player i interprets each active experience $[\phi_i; h_i]$ by associating it with a state $x = (g, \delta, y) \in X$ induced by her own deviation. That is, the player explains each active experience by the player's own unilateral deviation.

Finally, the coherence condition for $\mathcal{M}$ with $P_i(\sigma^*)$ is formulated as follows:

Condition CP: For any $[\phi_i; h_i]$, $[\phi_i; h_i]$ is in $P_i(\sigma^*)$ if and only if there is a state $x = (g, \delta, y) \in X$ such that $g_{-j} = g_{-j}^0$ for some $j \neq i$ and $[\hat{o}_i(x); \hat{u}_i(x)] = [\phi_i; h_i]$ hold.

This states that player i interprets a passive experience $[\phi_i; h_i]$ by associating it with a state $x = (g, \delta, y) \in X$ induced by a deviation (g_j, δ_j) of player $j \neq i$ or by a change in y. That is, player i regards a passive experience as induced by some other (imaginary) player or caused by an exogenous variable y. Note that because of the asymmetry between CA and CP, player i cannot detect who (or what y) induced the passive experiences, but he is certain that he has induced his active experiences.

Next, we give the following definitions.

Definition 14.1 An individual model $\mathcal{M} = \langle \hat{N}, \hat{Z}, \hat{o}_i, \hat{u}_i; x^0, X \rangle$ is coherent with the active experiences $A_i(\sigma^*)$ of player i if CA holds. We say that $\mathcal{M}$ is coherent with the experiences $A_i(\sigma^*) \cup P_i(\sigma^*)$ if CA and CP hold.

It is possible that we consider the coherence with passive experiences only, but in the present analysis, we will focus on coherence with the active experiences $A_i(\sigma^*)$ and coherence with the entire experiences $A_i(\sigma^*) \cup P_i(\sigma^*)$.

The true-game model $\mathcal{TG}$ is coherent with the entire experiences $A_i(\sigma^*) \cup P_i(\sigma^*)$ with respect to $\sigma^* = (f^*, r^*)$. This is a trivial statement since $\langle \hat{N}, \hat{Z}, \hat{o}_i, \hat{u}_i \rangle$ are constructed from our knowledge of the festival game Γ and x^0 and X are constructed from $\sigma^* = (f^*, r^*)$. However, since player i has no knowledge of the game Γ except as described in Postulate 1, $\mathcal{TG}$ is no more than one of many candidates. In fact, it will be shown in Sect. 14.6.1 that the true-game model $\mathcal{TG}$ hardly satisfies the other criterion rationalizability under our postulates.

We give two theorems to illustrate the effects of the coherence conditions. The first theorem states that when an inductively stable stationary state σ^* is given, the player knows that the maximum utility is obtained at the stationary state over the states that can be induced by her own deviations. The proofs of the results are given in the appendices of this chapter.

Theorem 14.1 (Utility maximization for player i) *Let* $\sigma^* = (f^*, r^*)$ *be an inductively stable stationary state. If an individual model* $\mathcal{M} = \langle \hat{N}, \hat{Z}, \hat{o}_i, \hat{u}_i; x^0, X \rangle$ *is coherent with the active experiences* $A_i(\sigma^*)$, *then*

$$\hat{u}_i(x^0) \geq \hat{u}_i(x) \tag{14.1}$$

holds for all $x \in A_i(\sigma^*)$.

This theorem implies that an individual model coherent with the active experiences should satisfy utility maximization for the player. This theorem has the same spirit as Proposition 13.1 that is, it is a manifestation of the postulate of inductive decision making in model $\mathcal{M}$.

Conversely, if every player has developed a model that is coherent with the active experiences and satisfies utility maximization, the stationary state σ^* is inductively stable.

Theorem 14.2 *Consider a stationary state* σ^* *where every player* $i \in N$ *has a model coherent with his active experiences* $A_i(\sigma^*)$. *Then* σ^* *is inductively stable if and only if the model of each player satisfies utility maximization* (14.1) *for all* $x \in A_i(\sigma^*)$.

For the same reason as that for Theorem 14.1, we can prove that player i can infer the ethnicity configurations of all festivals in the stationary state from his active experiences.

14.3   Rationalization

As remarked in Sect. 14.2.1, the cognitive bases for the festival game Γ and an individual model $\mathcal{M} = \langle \hat{N}, \hat{Z}, \hat{o}_i, \hat{u}_i; x^0, X \rangle$ are different in that in the former, player i does not know her own utility function U_i as a function; however, in the latter, the player perceives $\hat{u}_i$ as a function. Knowing player i's own utility function can be regarded as tantamount to maximizing it. Theorem 14.2 states that the coherence requirement with the active experiences together with Postulate 3 implies utility maximization over the active experiences. Postulate 1 states that player i knows the availability of friendly and unfriendly actions in addition to the choice of a festival location. Typically, the player faces such a choice problem when an outsider comes to her festival. Then, the player reacts always following her reaction function r_i^* and has never experienced the other attitude because only a deviation by a single player is perceivable by Postulate 2(b). Since player i knows that there is an alternative choice for her attitude, we require that an individual model $\mathcal{M} = \langle \hat{N}, \hat{Z}, \hat{o}_i, \hat{u}_i; x^0, X \rangle$ should satisfy utility maximization over these alternative choices.

To reflect this idea, we have the following definition of rationalization.

Definition 14.2 Given a stationary state σ^* player i rationalizes her strategy in $\mathcal{M}$ at σ^* if for any $x = (g, \delta, y) \in X$ with $[\hat{o}_i(x), \hat{u}_i(x)] \in P_i(\sigma^*)$ and for any $x' = (g', \delta', y') \in \hat{Z}$ with $g' = g$ and $\delta'_{-i} = \delta_{-i}$,

$$\hat{u}_i(x) \geq \hat{u}_i(x'). \tag{14.2}$$

This means that the prescribed reaction of player i against a deviation of some player maximizes utility function $\hat{u}_i$ in the model $\mathcal{M}$. The determination of the utility value $\hat{u}_i(x') = \hat{u}_i(g, (\delta_{-i}, \delta'_i, y')$ is a speculation in the sense that player i has never experienced $\hat{u}_i(x')$ in the past. Hence, player i can manipulate the model so that it satisfies (14.2). In our context of the festival game, it suffices to consider only the above case in addition to the choice of a festival location, which Theorem 14.1 addresses.

The term rationalization is motivated by the fact that player i adjusts her utility function $\hat{u}_i$ so that her behavior satisfies utility maximization. In the next section, we will see that rationalization plays a critical role in deriving prejudices, and we will show in Sect. 14.6.1 that the true-game model $\mathcal{TG}$ is hardly rationalizable.

14.4   Prejudicial Models

This section introduces hedonistic models, which explain the experiences in terms of the observables for an individual player. The class of hedonistic models is further divided into the two subclasses of naive hedonistic models and sophisticated hedonistic models. In a naive hedonistic model, one's utility is determined by one's own actions (f_i, δ_i) and an exogenous variable y. A model of this type can fully explain the active experiences. To explain the passive experiences, however, the model relies heavily on the exogenous variable y; otherwise, it is not rationalizable. On the other hand, a sophisticated hedonistic model allows its utility function to depend on the observed ethnicity configuration. A model of this type is coherent with all the experiences and rationalizable with a slight use of y. These models exhibit perceptual prejudices, and the latter additionally exhibits preferential prejudices.

We call an individual model $\mathcal{NH} = \langle \hat{N}, \hat{Z}, \hat{o}_i, \hat{u}_i; x^0, X \rangle$ a *naive hedonistic model* if its utility function $\hat{u}_i$ depends only on the player's actions and the exogenous variable y, i.e., it is expressed as

$$\text{NH4}: \quad \hat{u}_i(x) = \hat{u}_i(g_i, \delta_i, y) \text{ for all } x = (g, \delta, y) \in \hat{Z}.$$

This means that player i explains his observed utilities by his choices of a location and a friendly or unfriendly action together with the exogenous variable y.

If the utility function $\hat{u}_i$ of an individual model depends further on the observed ethnicity configuration, i.e., it is expressed as

$$\text{SH4}: \quad \hat{u}_i(x) = \hat{u}_i(g_i, \delta_i, \hat{E}_i(g), y) \text{ for all } x = (g, \delta, y) \in \hat{Z},$$

then the model $\langle \hat{N}, \hat{Z}, \hat{o}_i, \hat{u}_i; x^0, X \rangle$ is called a *sophisticated hedonistic model*, which we denote by $\mathcal{SH}$.

A sophisticated hedonistic model $\mathcal{SH}$ differs from a naive model $\mathcal{NH}$ in that in $\mathcal{SH}$ the observations are fully used to define the imaginary utility function $\hat{u}_i$, whereas in $\mathcal{NH}$ only his own choices are used. Since an individual player has, by Postulate 1, no knowledge about the structure of the society except these observables, it would be "natural" to construct a model based on these observables. This is in contrast with the true-game model $\mathcal{TG}$ in which the determination of the utility function needs more speculations on the structure of the society than in the hedonistic models.

When player i restricts his attentions to the active experiences $A_i(\sigma^*)$, he succeeds in constructing a naive hedonistic model without using any exogenous variable y, which is stated by the following theorem. Its proof is omitted.

Theorem 14.3 *Let* $\sigma^* = (f^*, r^*)$ *be an inductively stable stationary state. Then there is a naive hedonistic model* $\mathcal{NH} = \langle \hat{N}, \hat{Z}, \hat{o}_i, \hat{u}_i; x^0, X \rangle$ *coherent with the active experiences* $A_i(\sigma^*)$ *and such that* $\hat{o}_i$ *and* $\hat{u}_i$ *are independent of the exogenous variable* y, *i.e.,* $\hat{Z} = \{1, \ldots, \ell\}^{\hat{N}} \times \{(\smile), (\frown)\}^{\hat{N}}$.

Thus, the player succeeds in explaining his active experiences by ascribing his observed utilities to his actions. This explanation is fallacious from the objective point of view: player i finds an explanation of his observations based on an incorrect causal relationship, but it is still consistent with his observations. In this sense, the naive hedonistic model exhibits perceptual prejudices.

When player i also takes the passive experiences into account, a naive hedonistic model typically does not work well. In a segregation equilibrium, if the players in the smallest festival are all non-discriminators, a naive hedonistic model does work with a slight use of the exogenous variable y. In a larger festival, however, a discriminator cannot rationalize her behavior, and a non-discriminator needs to heavily use the exogenous variable to explain his passive experiences. The proof of the following theorem is given in the Appendix.

Theorem 14.4 *Let $\sigma^* = (f^*, r^*)$ be an inductively stable stationary state satisfying condition FA: $[r_i^*(f^*) = (\smile)$ for any $i \in N]$.*

(i) *Failure of Rationalization: Let player i be a discriminator against some ethnicity. If a naive hedonistic model $\mathcal{NH}$ is coherent with his experiences $A_i(\sigma^*) \cup P_i(\sigma^*)$, then it is not rationalizable at σ^*.*

(ii) *Heavy Dependence on y: Let player i be a non-discriminator toward any ethnicity. If a naive hedonistic model $\mathcal{NH}$ is coherent with the experiences $A_i(\sigma^*) \cup P_i(\sigma^*)$ and is rationalizable at σ^*, then*

$$\hat{u}_i(g_i^0, (\frown), y^0) \le \hat{u}_i(g_i^0, \delta_i^0, y) \le \min_{j \in N} U_j(\sigma^*) \tag{14.3}$$

holds for some y.

In (i), it also follows from coherence with $A_i(\sigma^*) \cup P_i(\sigma^*)$ that

$$\hat{u}_i(g_i^0, (\frown), y) \le \min_{j \in N} U_j(\sigma^*)$$

holds for some y. The point of (ii) is that the utility decrease in (14.2) is caused by a change in y, but in (i), it may be caused by a change in δ_i. If i is in the smallest festival, (14.3) would not restrict $\hat{u}_i$ because the right-hand side, $\min_{j \in N} U_j(\sigma^*)$, is the stationary utility value of the smallest festival. However, if player i is in a larger festival, his imaginary utility function $\hat{u}_i$ takes a value smaller than or equal to $\min_{j \in N} U_j(\sigma^*)$. Thus, typically, $\hat{u}_i$ heavily depends on the exogenous variable y.

Let us consider why a naive hedonistic model for, say, player i in C in festival 3 is not rationalizable nor has a heavy dependence on y (see Fig. 13.7). Suppose that player j in group A in festival 1 comes to festival 3. Then the number of friendly players of festival 3 after the trial ought to be less than or equal to that of festival 1 in the the stationary state; otherwise, player j would stay in festival 3. There are two cases to be considered: player i is a discriminator against ethnicity A or she is a non-discriminator.

If player i is a discriminator against A, she takes the unfriendly action in response to the presence of player j, though she takes the friendly action in the stationary state. In a naive hedonistic model, however, player i cannot justify this switch in behavior because the imaginary utility function $\hat{u}_i$ does not depend on ethnicity configurations. Consequently, player i fails to rationalize her own discriminatory behavior.

On the other hand, if player i is a non-discriminator toward ethnic group A, she does not switch her behavior, but the player remains friendly and observes a decrease in utility when some player in A is present in festival 3. In this case, player i can explain this change in utility only if she allows $\hat{u}_i$ to depend heavily on y.

When player i is a member in the smallest festival and all of its members are non-discriminators against any ethnicity, a naive hedonistic model works well with a slight use of the exogenous variable; $|Y| = 3$ and

$$\left| \hat{u}_i(g_i^0, \delta_i^0, y^0) - \hat{u}_i(g_i^0, \delta_i^0, y) \right| = 1$$

hold for all $y \in Y \setminus \{y^0\}$. The exogenous variable y is used to explain the changes induced when an insider goes out and when an outsider enters.

Given the preceding theorem and arguments, a naive hedonistic model does not work well for a player i in a large festival if the player takes all the passive experiences into account. However, if player i notices that utility changes caused by passive experiences are almost always associated with the presence of a different ethnicity, then the player could find that a sophisticated hedonistic model may be more suitable than a native hedonistic model. Indeed, whichever the player's festival may be and whichever her attitude may be, the player can succeed in constructing a sophisticated hedonistic model that is rationalizable and has only slight dependence on the exogenous variable y. The proof of the following theorem is given in the Appendix.

Theorem 14.5 (Hedonistic sophistication) *Let σ^* be an inductively stable stationary state satisfying condition FA. Then any player i has a coherent and rationalizable sophisticated hedonistic model $\mathcal{SH}$ with $Y = \{y^0, y^1\}$ and*

$$\hat{u}_i(g_i^0, \delta_i^0, E_i(g^0), y^0) - \hat{u}_i(g_i^0, \delta_i^0, E_i(g^0), y^1) = 1. \tag{14.4}$$

Perceptual prejudices are involved in a sophisticated hedonistic model $\mathcal{SH}$, as in a naive hedonistic model $\mathcal{NH}$, in that it involves a fallacious causal relationship. In $\mathcal{SH}$, the utility function $\hat{u}_i$ of player i further depends on the ethnicity configuration. In fact, we can regard this dependence as exhibiting that player i develops preferential prejudices against the outsiders' ethnicity. The greater the number of his fellow players who react as discriminators, the stronger these prejudices become. To look closely at this fact, let $\mathcal{SH}$ be a coherent and rationalizable hedonistic model satisfying (14.4).

Suppose that players i and j go to different festivals in the stationary state in a segregation equilibrium, and that festival f_j^* is smaller than festival f_i^*, i.e., the number of those who choose the same location as player j is smaller than the number of those who choose the same location as player i. In order to consider a deviation by player j, let f_j be the location that player j chooses upon deviation. We consider the case where $f_j = f_i^*$ holds. Then it must be the case that player j's utility does not increase even if she takes the friendly action, i.e.,

$$U_j(\sigma_{-j}^*, (f_j, (\smile))) \leq U_j(\sigma^*).$$

In response to the presence of player j, player i takes either the non-discriminatory action or the discriminatory action. Consider the cases one by one.

Case N: If player i is a non-discriminator, her utility induced by player j's deviation is equal to that of player j. This implies that player i's utility decreases significantly. Therefore,

$$\hat{u}_i(g_i^0, (\smile), \mathcal{E}_i(\sigma^*), y^0) = U_i(\sigma^*) > U_i(\sigma_{-j}^*, (f_j, (\smile)))$$
$$= \hat{u}_i(g_i^0, (\smile), \mathcal{E}_i(\sigma^*) \cup \{e(j)\}, y)$$
$$\geq \hat{u}_i(g_i^0, (\frown), \mathcal{E}_i(\sigma^*) \cup \{e(j)\}, y')$$

holds for all y'. The last inequality is due to rationalization. Thus, player i's utility necessarily decreases in the presence of $e(j)$. Player i still behaves in the friendly manner to ethnic group $e(j)$, but the player attaches a significant disutility value to ethnicity $e(j)$.

Case D: If player i is a discriminator, her utility induced by player j's deviation is zero. This implies

$$\hat{u}_i(g_i^0, (\frown), \mathcal{E}_i(\sigma^*) \cup \{e(j)\}, y) = 0$$

for some y by CP. Thus, we have

$$\hat{u}_i(g_i^0, (\smile), \mathcal{E}_i(\sigma^*), y^0) > 0 = \hat{u}_i(g_i^0, (\frown), \mathcal{E}_i(\sigma^*) \cup \{e(j)\}, y)$$
$$\geq \hat{u}_i(g_i^0, (\smile), \mathcal{E}_i(\sigma^*) \cup \{e(j)\}, y')$$

holds for all y'. The last inequality is due to rationalization.

Here player i behaves in the unfriendly manner in response to ethnicity $e(j)$ and receives zero utility, but this is still better than taking the friendly action.

In either case, a coherent and rationalizable $\mathcal{SH}$ exhibits prejudices against ethnicity $e(j)$. It may be slightly paradoxical to have the possibility that both discriminators and non-discriminators have preferential prejudices against some ethnicity, and the implications of this are discussed in Sect. 14.6. Here we note the following difference between the above two cases: if player i is a discriminator, she attaches zero or less payoff to the presence of ethnicity $e(j)$, which may be constant over any outside ethnicity the player discriminates against. On the other hand, if player i is a non-discriminator, then, typically, the smaller festival f_j^* is, the smaller the utility induced by the deviation. For example, if f_j^* is the smallest, the induced utility is not more than the minimum utility in the stationary state.

14.5 Escaping from Prejudicial Models

As we have seen, in our setup, segregation is sustained as an equilibrium phenomenon. Is there any measure to escape from such a segregation equilibrium?

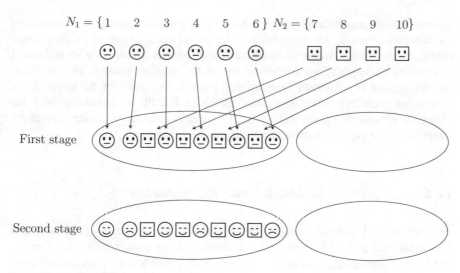

Fig. 14.1 Making prejudices incoherent

First, in order to make a prejudicial model, e.g., a sophisticated hedonistic model, incoherent with an experience, it is essential to have different experiences from the present one. In particular, we would like to have

$$\hat{u}_i(g_i^0, (\smile), \mathcal{E}_i(\sigma^*), y^0) \le \hat{u}_i(g_i^0, (\smile), \mathcal{E}_i(\sigma^*) \cup \{e_k\}, y')$$

for some y', i.e., the addition of some ethnicity e_k would not decrease player i's payoff in her model.

Once again, consider the previous example with $N_1 = \{1, \ldots, 6\}$ and $N_2 = \{7, \ldots, 10\}$. Figure 13.4 shows the stationary equilibrium behavior in the segregation equilibrium we analyzed. See Fig. 14.1. Suppose that this figure corresponds to a new experience, which is neither active nor passive. In fact, in this particular experience, all the players in N_2 go to location 1 and take the friendly action, while some players in N_1 also take the friendly action. In this experience, the number of friendly players is seven, and therefore, the players in N_1 who take the friendly action will obtain a higher payoff than the payoff in the stationary state. To explain this experience, a player in N_1 has to throw away the prejudicial model that puts a lower payoff when $E = \{e_1, e_2\}$ than when $E = \{e_1\}$.

14.6 Discussion

This analysis presents a new theory, called inductive game theory, which targets the formation and emergence of individual views about society from experiences. Instead of general situations, we treat a specific game (the festival game) and specific object phenomena: discrimination and prejudices. This research strategy of considering specific problems was deliberate because the sound development of a theory needs simultaneous considerations of theory and applications. Here, we give additional discussions on implications of our theory on the existing game theory, examine the interactions of individual behavior and models, and consider the implications on relevant sociological and economics literature. Finally, we comment, from the viewpoint of inductive game theory, on the common knowledge assumption on the structure of a game in deductive game theory.

14.6.1 Sequential Rationality and Rationalization

Proposition 13.1 characterizes inductive stability to be Nash equilibrium, and Theorems 14.1 and 14.2 relate Nash equilibrium to coherence with the active experiences. Game theory literature provides another well-known concept: subgame perfect equilibrium (or sequential equilibrium, to be precise). As remarked at the end of Sect. 13.3, sequential rationality is not derived from experiences, in contrast with Nash equilibrium, because the player has no experience of the alternative attitude in the second stage of the festival. On the other hand, we have required rationalization for the second choice in an individual model. In the following, we consider the relationship between these concepts by using the true-game model $\mathcal{TG}$.

In the festival game Γ, sequential rationality on a strategy profile $\sigma^* = (f^*, r^*)$ is relevant only to the nodes in the second stage reachable by a unilateral deviation from σ^*. In the terminology of the present model, the relevant scope for subgame perfection is $\cup_i P_i(\sigma^*)$; each of these nodes is reached by a unilateral deviation of a player. We require sequential rationality over these nodes. We say that $\sigma^* = (f^*, r^*)$ satisfies *sequential rationality* (together with consistent beliefs) over $\cup_i P_i(\sigma^*)$ if for all $i, j \in N$ with $f_j^* \neq f_j = f_i^*$ and for $\delta_i, \delta_j = (\smile), (\frown)$,

$$U_i(\sigma^*_{-j}, (f_j, \delta_j)) \geq U_i(\sigma^*_{-i,j}, (f_i^*, \delta_i), (f_j, \delta_j))$$

holds, where $(\sigma^*_{-i,j}, (f_i^*, \delta_i), (f_j, \delta_j))$ is obtained from σ^* by replacing σ_i^* and σ_j^* with (f_i^*, δ_i) and (f_j, δ_j). This means that if an outsider j comes to $f_j = f_i^*$, the prescribed action $r_i^*(f_{-j}^*, f_j)$ of player i maximizes her payoff.[1]

Although there always exists a coherent and rationalizable sophisticated hedonistic model, this may not be the case for the true-game model. Indeed, a striking result is that the true-game model typically fails to be rationalizable, as shown in the following theorem, whose proof is given in the Appendix.

---

[1] Note that we ignore the deviations by an insider. For example, if the festival f_i^* has only one player of some ethnicity and if this player goes out, player i would observe a change in the ethnicity configuration. The above definition does not take this case into account. However, it was proved in Sect. 13.4 that we do not have this case in a fully active equilibrium.

Theorem 14.6 (Rationalization for the true-game model) *Let σ^* be an inductively stable stationary state satisfying condition FA. Then the true-game model TG of player i in σ^* is rationalizable for all $i \in N$ if and only if σ^* satisfies sequential rationality over $\cup_i P_i(\sigma^*)$.*

Theorem 14.6 states that rationalization corresponds to sequential rationality. As discussed in Sect. 13.3, sequential rationality is not guaranteed under our postulates, and also, Theorem 13.2 implies that there are many Nash equilibria that do not satisfy sequential rationality. Therefore, rationalization typically rejects the true-game model.

If we do not necessarily regard sequential rationality as a requirement for the original objective game, but apply it to the individual models, then sequential rationality and rationalization may be regarded as conceptually equivalent. From this point of view, when an individual player thinks about the society as a model, she is inclined to assume sequential rationality (rationalization), but unless the player has enough experience of the payoff function, the player may also be inclined to adjust (rationalize) her payoff function so as to satisfy sequential rationality. This is in sharp contrast with the derivation from Nash equilibrium in the sense of Proposition 13.1 and Theorem 14.2.

14.6.2 Interactions Between Models and Behavior

It may be already clear from our discussions that coherence and rationalization express different reasoning processes. Coherence of a model is attained through the process of induction based on experiences, while rationalization is obtained through the introspection of consistency between the model and behavior. When the model fails to be rationalizable, the player either alters the interpretation of the model or changes her behavior. This subsection discusses their evolution in a dynamic context, considering some possible scenarios.

As the reference point of induction, we recall the mere-enumeration model mentioned in Remark 1 at the end of Sect. 14.2. The mere-enumeration model does not go much beyond the state of collecting the experiences $A_i(\sigma^*) \cup P_i(\sigma^*)$ since it gives no causal relationship between the player's observations and satisfaction. In this sense, it represents the state of mind of the player who has had experiences, is conscious of them, but has not deliberated on them further. In contrast, the true-game model could be regarded as the ultimate goal from the objective point of view. The problem associated with it is whether or not an individual player needs to (or is able to) consider the true-game model after the full deliberation of her experiences.

Suppose that player i has experiences $A_i(\sigma^*) \cup P_i(\sigma^*)$, and that the player is conscious of these experiences. If the player wants to have a better explanation of her observations including utility values, she may start thinking about hedonistic models.

First, consider naive hedonistic models. It follows from Theorem 14.3 that if player i cares only about active experiences or if the player is in the smallest festival where every player is a non-discriminator, a naive hedonistic model could work and there is no need to think about the society any further. Let player i be in a larger festival, and suppose that he takes passive experiences as well as active ones into account and introspects his explanation of experiences. Theorem 14.4 (i) states that if one is a discriminator against some ethnic group, one cannot rationalize one's own behavior in a naive hedonistic model. Hence a discriminator would sophisticate his explanation, and may reach a sophisticated hedonistic model. Theorem 14.4 (ii) states that if one is a non-discriminator toward any ethnic group, one can succeed in explaining his behavior in a naive hedonistic model by using an exogenous variable. However, if one wants a better explanation to avoid a heavy use of an exogenous variable, one may sophisticate one's model. In either case, a natural candidate for a modification would be a sophisticated hedonistic model.

When the players reach sophisticated hedonistic models that are coherent with experiences and rationalizable, no further change will be induced. Then the inductively stationary state is truly stable. In this process, models have deviated from naive hedonistic models to become models whose utility functions involve ethnic configurations as a fallacious explanatory variable. This is the emergence of preferential prejudices against an ethnic group.

However, there is another logical possibility: many discriminators change their actions to the friendly ones at the same time, although it could be rather accidental and may hardly ever occur. However, if this happens, their utility values would not decrease even when an outsider from a smaller festival visits their festival. Since such an outsider receives a higher utility value, he would stay in the larger festival. The dissolution of segregation would then follow, but this possibility would be very accidental.

Finally, let us look at what happens if player i starts seeking the true-game model in the process of deviating from a naive hedonistic model or finding a better explanation of her experiences than a sophisticated hedonistic model. Suppose that the player thinks about the true-game model. The player must be uncertain about the correctness of her true-game model because the player has no evidence other than her experiences (therefore, it must be very difficult for him to think about the true-game model).

From Theorem 13.2, a segregated equilibrium typically does not satisfy sequential rationality, and Theorem 14.6 implies that the true-game models of the players in the larger festival are not rationalizable. Since the player is uncertain about her model, she modifies it so as to rationalize her behavior. One possibility is to go to or return to a sophisticated hedonistic model. Thus, a sophisticated hedonistic model is also regarded as stable in this sense.

When players have reached coherent and rationalizable sophisticated hedonistic models, they cannot reject such prejudicial models unless the players have new experiences, for example, by going to another society with a different stationary state.

14.6.3 Comparisons with Merton's Classification

Finally, we mention some implications to the studies of discrimination and prejudices in sociology and economics.

Merton (1949) suggested four ideal types by combining the propensity for discriminatory behavior with the presence of prejudice.

These types have counterparts in our theory. To begin with, we have two categories in terms of behavior. On one hand, non-discriminators are the ones who do not take different actions when someone from another ethnic group comes to their location. On the other hand, discriminators are the ones who take different actions, typically unfriendly actions rather than friendly actions, when someone from another ethnic group comes to their location.

Next, there are two categories, unprejudiced and prejudiced. On one hand, we interpret "unprejudiced" as the players whose utility functions in their models are independent of ethnicity configurations. Examples are the true-game model and naive hedonistic models. On the other hand, we interpret "prejudiced" as those whose utility functions in their models depend on ethnic configurations. In our analysis, those who have a sophisticated hedonistic model are considered "prejudiced".

These classifications give us four combinations as in Table 14.1. First, "all-weather liberals" are the non-discriminators without prejudice. Their utility functions in their models are independent of ethnicity configurations, and they do not take discriminatory behavior against players in different ethnic groups.

Second, "active bigots" are the opposite of "all-weather liberals." They discriminate against some ethnic groups, and they have prejudice in the sense that their utility functions in their models involve the ethnicity configurations in a critical manner: if someone from some particular ethnic groups comes to their festival, their utility value in the model decreases.

Third, "timid bigots" are the non-discriminators, but their utility functions in their models are based on negative images of other ethnic groups.

Fourth, "fair-weather liberals" are the discriminators but explain their utilities without referring to ethnicity configurations.

Merton (1949) introduced those four types to examine the causal relationship between prejudices and discrimination. If prejudices induce discrimination, then people could simply be categorized into either all-weather liberals or active bigots. However, Merton argued (also see Marger 1991, Chap. 3, for recent assessments of this view) that because all four types of people are observed in our society, the causal relationship from prejudices to discriminatory behavior is questionable. In our

Table 14.1 Merton's classification

	Unprejudiced	Prejudiced
Non-discriminators	All-weather liberals	Timid bigots
Discriminators	Fair-weather liberals	Active bigots

theory, as discussed above, prejudices may emerge in evolutions of the behavior of players together with their views on the society, and those four types of people seem to appear. In this sense, our theory supports Merton's view, although our theory goes beyond his.

In neoclassical economics (e.g., Becker 1957), it has been assumed that behavioral attitudes are determined by mental attitudes. Thus, the view described in this analysis looks contradictory to neoclassical economics. It should be noticed, however, that our theory is about a long-run situation, and that if we take a snapshot of this long-run situation, the causal relation would become one-directional; that is, prejudices induce discriminatory behavior.

14.6.4   *Large Societal Games Versus Small Micro Games*

If the source of the individual knowledge of the structure of a game is the inductive construction of a view about the game from experiences, then deductive game theory could be regarded as partly a result of inductive game theory. In deductive game theory, it is a traditional view that the structure of the game is common knowledge. On the other hand, we have considered an individual model that has only an imaginary utility function of the player. This assumption can be easily extended to a social model including the utility functions of other imaginary players. Nevertheless, it would be a different problem whether these utility functions are assumed to be common knowledge. It is important to notice that common knowledge requires each player to look into the other players' minds. In our context, since the individual player does not know even the identities of the other players, this requirement could hardly be met. Deductive game theory with the common knowledge assumption may appear to have no place in inductive game theory.

We take, however, a view to regard deductive game theory with the common knowledge assumption as an ideal (limit) case in a direction, different from the one taken in this analysis, in inductive game theory. The basic criterion is whether or not an individual player looks into the other players' minds and thinks about what the other players are thinking. This criterion may be regarded as being met in a small micro game where the players are playing the game face-to-face. This is another direction of inductive game theory, and Kaneko (1998) considers that the common knowledge assumption is regarded as a limit case of the evolution of the knowledge structure obtained by induction and deduction in such a small micro game.

In a large society, the assumption of looking into the minds of others is simply inadequate because even knowing the identities of other people is difficult. After all, what we have discussed is a problem of a large societal game. An individual model is an external description of the society, and should be interpreted as a partial description of the society, although our definition may still be too detailed.

Appendix

Before we prove Theorem 14.1, we present the following lemma.

Lemma 14.7 *Suppose that $\mathcal{M}$ is coherent with $A_i(\sigma^*)$. Let $x = (g, \delta, y)$ and $x' = (g', \delta', y')$ in X satisfy $g = g' = (g^0_{-i}, g_i)$, $(g_i, \delta_i) \neq (g^0_i, \delta^0_i)$, and $\delta_i = \delta'_i$. Then $[\hat{o}_i(x), \hat{u}_i(x)] = [\hat{o}_i(x'), \hat{u}_i(x')]$.*

Proof By the if part of CA, both $[\hat{o}_i(x), \hat{u}_i(x)]$ and $[\hat{o}_i(x'), \hat{u}_i(x')]$ are in $A_i(\sigma^*)$. If two experiences $[\phi_i; h_i]$ and $[\phi'_i; h'_i]$ in $A_i(\sigma^*)$ are induced by the same deviation (g_i, δ_i) from $\sigma^* = (f^*, r^*)$, then they coincide, since they are both expressed as $[g_i, \delta_i, \mathcal{E}_i(f^*_{-i}, g_i); U_i(\sigma^*_{-i}, (f_i, \delta_i))]$. Since the way player i deviates in x is the same as that in x', so are $[\hat{o}_i(x), \hat{u}_i(x)]$ and $[\hat{o}_i(x'), \hat{u}_i(x')]$. $\qquad\square$

Proof of Theorem 14.1 First, we have $\hat{u}_i(x^0) = U_i(\sigma^*)$ by CS. Consider $(\sigma^*_{-i}, (g_i, \delta_i))$. This gives an active experience $\phi_i = (g_i, \delta_i, \mathcal{E}_i(\sigma^*_{-i}, (g_i, \delta_i)))$ and $h_i = U_i(\sigma^*_{-i}, (g_i, \delta_i))$. By CA, there is a state $x' = ((g^0_{-i}, g_i), (\delta'_{-i}, \delta_i), y') \in X$ such that $\hat{o}_i(x') = \phi_i$ and $\hat{u}_i(x') = h_i$. Now, we take an arbitrary $x = ((g^0_{-i}, g_i), (\delta_{-i}, \delta_i), y) \in X$. Lemma 14.7 implies $h_i = \hat{u}_i(x') = \hat{u}_i(x)$. Since σ^* is inductively stable, we have $U_i(\sigma^*) \geq h_i$ by Proposition 13.1. Hence, $\hat{u}_i(x^0) = U_i(\sigma^*) \geq h_i = \hat{u}_i(x') = \hat{u}_i(x)$. $\qquad\square$

Proof of Theorem 14.2 The *only-if* part is Theorem 14.1. Thus, it suffices to show that if the model $\mathcal{M} = \langle \hat{N}, \hat{Z}, \hat{o}_i, \hat{u}_i; x^0, X \rangle$ of player i is coherent with $A_i(\sigma^*)$ and satisfies inequality (14.1), then his stationary actions maximize her objective payoff function. Consider an arbitrary (f_i, δ_i). This induces an experience $[\phi_i; h_i] = [f_i, \delta_i, \mathcal{E}_i(f^*_{-i}, f_i); U_i(\sigma^*_{-i}, (f_i, \delta_i))] \in A_i(\sigma^*)$. Since $\mathcal{M}$ is coherent with active experiences $A_i(\sigma^*)$, there is a state $x = (g, \delta, y) \in X$ such that $g_{-i} = g^0_{-i}$, $\hat{o}_i(x) = (f_i, \delta_i, \mathcal{E}_i(f^*_{-i}, f_i))$, and $\hat{u}_i(x) = U_i(\sigma^*_{-i}, (f_i, \delta_i))$. Also, $\hat{u}_i(x^0) = U_i(\sigma^*)$ by CS. Then $U_i(\sigma^*) = \hat{u}_i(x^0) \geq \hat{u}_i(x) = U_i(\sigma^*_{-i}, (f_i, \delta_i))$. $\qquad\square$

Proof of Theorem 14.4 For Part (i), let player j come to the festival k of player i with the friendly action. Suppose that player i takes the unfriendly action to j's presence. Then $U_i(\sigma^*_{-j}, (f_j, (\smile))) = 0 < U_i(\sigma^*)$. By CP, $\hat{u}_i(g, \delta, y) = \hat{u}_i(g^0, (\frown), y) = 0$ holds for some (g, δ, y). However, since $\hat{u}_i(g^0_i, (\smile), y^0) = U_i(\sigma^*) > 0$ by CS, we have

$$\hat{u}_i(g, \delta, y) = \hat{u}_i(g^0_i, (\frown), y) = 0 < \hat{u}_i(g^0_i, (\smile), y^0) = \hat{u}_i(g, (\delta_{-i}, (\smile)), y^0).$$

This violates rationalization.

For Part (ii), let player j be a player in the smallest festival. Then his payoff $U_j(\sigma^*)$ is the lowest. When j comes to $f_j = f^*_i$ and take the friendly action, $U_j(\sigma^*_{-j}, (f_j, (\smile))) \leq U_j(\sigma^*)$ holds; otherwise, j would stay at f_j. By CP, there is (g, δ, y) such that $g_i = g^0_i$, $\delta_i = r^*_i(f^*_{-j}, f_j) = (\smile) = \delta^0_i$, and

$$\hat{u}_i(g^0_i, (\smile), y) = U_i(\sigma^*_{-j}, (f_j, (\smile))) = U_j(\sigma^*_{-j}, (f_j, (\smile)))$$

hold. Thus, $\hat{u}_i(g_i^0, \delta_i^0, y) \leq U_j(\sigma^*) = \min_{k \in N} U_k(\sigma^*)$. By rationalization, we have

$$\hat{u}_i(g_i^0, (\smallfrown), y^0) \leq \hat{u}_i(g_i^0, \delta_i^0, y) \leq U_j(\sigma^*) = \min_{k \in N} U_k(\sigma^*).$$

$\square$

Proof of Theorem 14.5 Let $\sigma^* = (f^*, r^*)$ be an inductively stable stationary state satisfying FA. Let $f_j = f_i^*$ for player j below. We define the constituents of a sophisticated hedonistic model other than utility function $\hat{u}_i$ as follows:

SH1 $\hat{N}$ is an arbitrary imaginary player set partitioned into nonempty disjoint ethnic groups $\hat{N}_1, \ldots, \hat{N}_{e_0}$ with $i \in \hat{N}_{e(i)}$ and $|\hat{N}_{e(i)}| \geq 2$;

SH2 $\hat{Z} = \{1, \ldots, \ell\}^{\hat{N}} \times \{(\smile), (\smallfrown)\}^{\hat{N}} \times \{-1, 0\}$;

SH3 $\hat{o}_i(x) = (g_i, \delta_i, \hat{E}_i(g))$ for all $x = (g, \delta, y) \in \hat{Z}$;

SH5 $x^0 = (g^0, \delta^0, 0)$, where $\delta^0 = ((\smile), \ldots, (\smile))$ and $g^0 = (g_j^0)_{j \in \hat{N}}$ is defined by:

for each $j \in \hat{N}_e$ ($e = 1, \ldots, e_0$), $g_j^0 = f_k^*$ for some $k \in N_e$;

SH6 $X = \{x^0\} \cup X_A \cup X_P$, where

$$X_A = \left\{((g_{-i}^0, g_i), (\delta_{-i}^0, \delta_i), 0)\right\}_{(g_i, \delta_i)}$$

$$X_P = \{((g_{-j}^0, f_j), (\delta_{-i}^0, r_i^*(f_j, \hat{E}_i(g_{-j}^0, f_j))), 0))\} \cup \{(g^0, \delta^0, -1)\}.$$

We define a utility function $\hat{u}_i : \{1, \ldots, \ell\} \times \{(\smile), (\smallfrown)\} \times 2^{\{1, \ldots, e_0\}} \times \{-1, 0\} \to \mathbb{R}$ by

$$\hat{u}_i(g_i, \delta_i, E, y) = \begin{cases} \delta_i(\mu_i(\sigma_{-i}^*, (g_i, \delta_i)) + y) & \text{if } \mathcal{E} = \mathcal{E}_i(f_{-i}^*, g_i) \text{ for some } g_i, \\ \delta_i(\mu_i(\sigma_{-j}^*, (f_j, \delta_i)) + y) & \text{if } \mathcal{E} = \mathcal{E}_i(f_{-j}^*, f_j) \text{ for some } j, \delta_i = r_i^*(f_{-j}^*, f_j), \\ h^-(g_i, \delta_i, E, y) & \text{if } \mathcal{E} = \mathcal{E}_i(f_{-j}^*, f_j) \text{ for some } j, \delta_i \neq r_i^*(f_{-j}^*, f_j), \\ \text{arbitrary} & \text{otherwise,} \end{cases}$$

where $h^-(g_i, \delta_i, E, y)$ is a real number not greater than $U_i(\sigma_{-j}^*, (f_j, 0))$. FA and Theorem 13.2 guarantee that the function is well-defined. The first case gives utility values to i's own deviations, which together with the observation function $\hat{o}_i$ implies coherence with active experiences. This covers also the case where an insider takes the unfriendly action or moves out of festival f_i^*. The second case gives utility values to the situations where outsiders come to festival f_i^*, which together with $\hat{o}_i$ implies coherence with passive experiences. The third case gives utility values to the unexperienced cases where an outsider come to festival f_i^* but he took the action not prescribed by r_i^*. In fact, we set $h^-(g_i, \delta_i, E, y)$ so that the sophisticated hedonistic model is rationalizable. $\square$

Proof of Theorem 14.6 The if part is straightforward. We show the *only-if* part. By FA, σ^* is a fully active equilibrium. If the players in each festival are all non-discriminators or all discriminators toward each ethnicity of an outsider, then σ^* enjoys sequential rationality over $\cup_i P_i(\sigma^*)$. Now we show the contrapositive of the *only-if* part. Suppose that σ^* does not enjoy sequential rationality over $\cup_i P_i(\sigma^*)$.

Then some festival has non-discriminators as well as discriminators toward some ethnicity.

Let k be a festival where some are discriminators and some are non-discriminators when an outsider j comes to f_j. Let i and i' be a discriminator and a non-discriminator, respectively, in k toward the ethnicity of j. There are two cases to consider: (a) $\mu_i(\sigma^*_{-j}, (f_j, (\smile))) \geq m_0$; and (b) $\mu_i(\sigma^*_{-j}, (f_j, (\smile))) < m_0$.

In case (a), let x be the path determined by $(\sigma^*_{-j}, (f_j, (\smile)))$. Then $\hat{u}_i(x) = U_i(\sigma^*_{-j}, (f_j, (\smile))) = 0$. Let $\sigma'_i = \sigma'_j = (f_j, (\smile))$, $\sigma'_{-i,j} = \sigma^*_{-i,j}$, and x' be the path determined by σ'. Then, we have $0 < U_i(\sigma') = \hat{u}_i(x')$. Hence, $\mathcal{TG}$ of player i is not rationalizable.

Consider case (b) in which $\mu_{i'}(\sigma^*_{-j}, (f_j, (\smile))) < \mu_i(\sigma^*_{-j}, (f_j, (\smile))) < m_0$ holds. Define σ', the strategy profile given by $\sigma'_{i'} = (f_j, (\frown))$, $\sigma'_j = (f_j, (\smile))$, $\sigma'_{-i',j} = \sigma^*_{-i',j}$, and let x' be the path determined by σ'. Then $\hat{u}_{i'}(x) = U_{i'}(\sigma^*_{-j}, (f_j, (\smile))) < 0 = U_{i'}(\sigma') = \hat{u}_{i'}(x')$. Hence, $\mathcal{TG}$ of player i' is not rationalizable. $\qquad\square$

Chapter 15
Everyone on the Island Spoke Sign Language

15.1 Introduction

This chapter takes a game-theoretic approach to disability-related issues by constructing a model that studies the case of hereditary deafness on Martha's Vineyard Island, Massachusetts, USA. In the past centuries, where the island community successfully adjusted to the hereditary deafness and made it a "non-disability."[1]

The present analysis asserts that disability presents not only physical/medical issues, but societal/economic issues, and that the notion of disability is socially relative. It further proposes that game theory is useful for understanding disability-related issues such as prejudice, which in turn call for a new approach in game theory like inductive game theory.

Martha's Vineyard Island, situated off the Massachusetts coast of the USA, had a high frequency of hereditary deafness for more than 250 years. In many modern societies, those who use sign language are considered to have a disability and suffer from social exclusion. On this island, however, due to a significant population of deaf people, many, if not all, of those without hearing impairment used sign language for communication.

In order to study this case by using game theory, we should first identify the key fundamental difference between the group of persons with hearing impairment and the group of persons without impairment. The merit of using game theory to study disability is that it forces us to clarify the individual (physical) difference as opposed to societal difference between the two groups of people. Any societal difference, disadvantage, or disability is explained as an equilibrium phenomenon rather than something that is socially predetermined, implying that disability is often a socially relative concept.

In the case of Martha's Vineyard, we differentiate the two groups according to their selection of communication tool. The only fundamental difference between the two is that the former cannot choose oral language as a communication tool,

---

[1] This chapter is based on Matsui (2017).

© Springer Nature Singapore Pte Ltd. 2019
A. Matsui, *Economy and Disability*, Economy and Social Inclusion,
https://doi.org/10.1007/978-981-13-7623-8_15

while the latter can choose oral language, sign language, or both. This difference in individual traits induces two types of equilibria. In the first type of equilibrium, those who cannot choose oral language become persons with disability in the sense that they cannot communicate with the majority and thus cannot access the resources shared by the majority. In the second type of equilibrium, although they may still be a minority, they can access the resources.

The model and its analysis in the present chapter are roughly described as follows. In the beginning, there are two types of continua of agents, the deaf and the non-deaf. There are two stages. In the first stage, the non-deaf agents become either bilinguals or monolinguals. As we shall see later, this process is expressed as an evolutionary process rather than a rational decision making process.

In the second stage, agents are classified as deaf people, bilinguals, and monolinguals. They are randomly matched to form a trio to play a three-person bargaining game with language constraints. Each trio can produce one unit of good and share it upon agreement, which can be reached only if they can communicate.[2]

We consider two scenarios. The first one is a majority bargaining game in which an agreement can be reached if one responder agrees to a proposal. The second one is a unanimity bargaining game in which both responders must agree in order to reach an agreement. These two bargaining games induce qualitatively different outcomes. The first game exhibits strategic complementarity, while the second exhibits strategic substitutability; the more bilinguals there are, the more (resp. less) incentive a non-deaf agent has to become a bilingual in the majority (resp. unanimity) bargaining game.

We consider these two games *not* because they are the only possible games played in the community. People in the community must have played a variety of games. In language selection games, however, it is often taken for granted that the more people learn a certain language, the more incentive one has to learn this language. We show in the sequel that this assertion needs to be examined.

Following Chaps. 13 and 14, this analysis discerns three approaches in game theory, deductive, evolutionary, and inductive approaches (see Fig. 15.1). The deductive approach is based on the rationality hypothesis, assuming that players deduce their strategies from their knowledge of the game they play. The evolutionary approach typically assumes that players have no knowledge of the game they play. Players may not even "choose" actions but can act only according to some program. Here, the survival of the fittest may apply; the higher the payoff one obtains, the better chance one has of producing offsprings and the likelier one is to take this action in the future. The inductive approach assumes no (sufficient) prior knowledge of the game players play. Instead, like in evolutionary game theory, the players play the game to accumulate experiences. Unlike in evolutionary game theory, however, the players try to construct a model of the game based on their experiences, as described above.

---

[2]Three is the smallest number by which a majority and minority can emerge in the bargaining game.

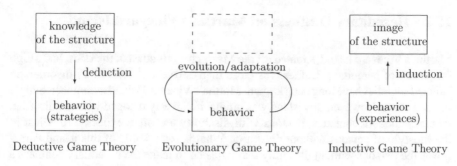

Fig. 15.1 Three approaches to game theory

The present analysis uses these three approaches in its analysis of hereditary deafness on Martha's Vineyard Island. The second-stage bargaining game is analyzed through the deductive approach, as agents often understand the structure of the game they play and make rational decisions in the bargaining process. The three-person bargaining game has been studied extensively (see, e.g., Okada 1996). The novelty of the present analysis is the incorporation of an additional structure of language selection in its analysis.

The first stage of language selection is analyzed through the evolutionary approach since language acquisition often occurs in childhood and is often affected by the past phenomena, such as the language used by their parents. This analysis uses the best response dynamics from Chap. 8, although any reasonable dynamics would suffice.[3] This part of the analysis is related to Chaps. 10 and 11 where persons with different traits interact with each other and adapt to their society in an evolutionary manner.

One of the major focuses of disability studies is prejudice against persons with disability. According to the Cambridge Dictionary online, prejudice is "an unfair and unreasonable opinion or feeling, especially when formed without enough thought or knowledge." By nature, prejudice involves some sort of wrong beliefs and limited reasoning. The deductive approach is thus unsuitable, as it begins with "correct" beliefs and "right" reasoning processes. The evolutionary approach is also inappropriate since its main engine of selection is the survival of the fittest rather than contemplation. Wrong beliefs are formed through (often limited) experiences and (again often inadequate) contemplation. This study therefore uses the inductive approach to see how prejudice emerges.

The rest of this chapter is organized as follows. Section 15.2 presents an example of hereditary deafness in which members of a society adapted to disability in an interesting way that motivated our study. Section 15.3 presents the framework of our analysis. Sections 15.4, 15.5, and 15.6 present an analysis of the model based on deduction, evolution, and induction, respectively. Section 15.7 concludes the chapter.

---

[3]See also Kandori (1997) for an extensive survey.

15.2   Hereditary Deafness on Martha's Vineyard Island

Martha's Vineyard Island, situated off the Massachusetts coast of the USA, had a high frequency of hereditary deafness for more than 250 years.[4] Hereditary disorders in isolated societies have long been known. Martha's Vineyard Island is another example of such cases. However, this island was unique in the way it coped with the disorder. In many modern societies, persons with disability are often expected to adjust to the lifestyle of persons without disability. What is remarkable of this island is that it was the persons without disability that adjusted to make the hereditary deafness a "non-disability".

In the nineteenth century, the frequency of deafness at birth was about 1 in 6000 in the USA. On the other hand, this number was 1 in 155 on Martha's Vineyard, and 1 in 25 in the town of Chilmark, which is located on the western part of the island. As a result, "Deafness was seen as something that just 'sometimes happened'; anyone could have a deaf child."[5] People's attitudes toward deafness are summarized by Groce (1985) as follows (p. 51):

- You'd never hardly know they were deaf and dumb. People up there got so used to them that they didn't take hardly any notice of them.[6]
- It was taken pretty much for granted. It was as if somebody had brown eyes and somebody else had blue. Well, not quite so much–but as if, somebody was lame and somebody had trouble with his wrist.
- They were just like anybody else. I wouldn't be overly kind because they, they'd be sensitive to that. I'd just treat them the way I treated anybody.

Being deaf is not a handicap per se. It is social isolation that creates a handicap. On Martha's Vineyard, this isolation hardly occurred as the islanders learned sign language in childhood. They needed to learn sign language "to communicate with deaf adults as well as deaf playmates" (Groce 1985, p. 54).

One may casually conclude that on the island, everyone used sign language. It should also be mentioned, however, that some people on the island did not use sign language. Indeed, Groce (1985) wrote about an informant who felt uncomfortable but did not learn the sign language. The informant said, "I used to feel chagrined because I couldn't speak the sign language" (Groce 1985, p. 56).

---

[4]This section is based on Groce (1985).

[5]Groce (1985), pp. 50–51.

[6]The term "deaf and dumb" now has pejorative connotations, but Groce (1985) retains this term as it is not pejorative on Martha's Vineyard. Therefore, it is retained in all quotes.

15.3   Model

We consider a society that consists of infinitely many agents with either one of two natural traits, deaf (D) and non-deaf (N). The fraction of the deaf agents is exogenously given by $\alpha \in (0, 1)$.

There are two stages in the game. In the first stage, N type agents simultaneously choose N_b or N_m, where b stands for "bilingual", and m for "monolingual". To become an N_b agent, an N agent incurs cost $d > 0$.

In the second stage, the agents are randomly matched to play a three-person bargaining game. Each agent in this stage has one of three statuses, D, N_b, and N_m. The deaf agents can use a sign language, while the non-deaf agents can use a spoken language. Among N type agents, those who become an N_b agent can use both sign language and spoken language, while those who become an N_m agent can use only spoken language.

Prior to the bargaining stage, there is an additional substage where type N_b agents can decide which language to use in the bargaining game, sign language only, oral language only, or both languages. An N_b agent who decides to use the both languages has to make the same offer in the both languages.

Each bargaining game is played by three agents: call them 1, 2, and 3. A bargaining game is infinitely repeated until an agreement is reached. In each period, one of the three agents is randomly chosen to be a proposer. Suppose Agent 1 is chosen as a proposer. Agent 1's proposal is denoted by $x = (x_1, x_2, x_3)$ with $x_1 + x_2 + x_3 = 1$. After x is proposed, the other two agents choose A ("accept") or R ("reject") on condition that they understand the language used by the proposer. Similarly, we denote by y (resp. z) the offer proposed by agent 2 (resp. 3).

If a responder does not use the language of the proposer, then the responder has to choose R. The actual game tree, therefore, depends upon the profile of the agents' statuses. We write $s = (s_1, s_2, s_3)$ $(s_i \in \{D, N_b, N_m\}, i = 1, 2, 3)$ to denote the status configuration of the bargaining game. Also, we sometimes write like (D_1, D_2, N_b) when we would like to pay attention to the identity of agents therein.

The game ends if the agents reach an agreement. Suppose that x is offered, and the game ends in the tth period. Then Agent i's payoff is given by $\delta^{t-1} x_i$, where $\delta \in (0, 1)$ is a common discount factor. If no agreement is reached in any period, then the payoff of each agent will be zero.

The analysis of this chapter considers two classes of bargaining games, the majority bargaining games and the unanimity bargaining games with language constraints.

15.3.1   The Majority Bargaining Games

In the majority bargaining game with language constraints, at least two agents (the majority) need to reach an agreement. To be precise, if at least one responder chooses "A", then the game ends, and x is realized as the outcome of the bargaining game. The

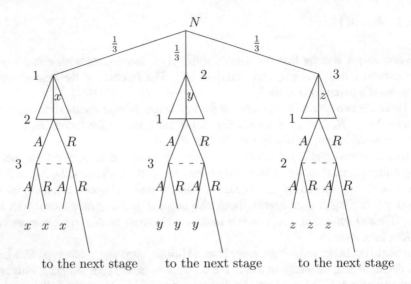

Fig. 15.2 A majority bargaining stage when everyone understands the language used in the bargaining

bargaining stage at which everyone understands the language used in the bargaining stage is shown in Fig. 15.2.

If the type profile is different, we have a different game tree. The rule for constructing a corresponding game is as follows. If an agent does not understand the language of the proposer, then this agent's response has to be R, and to express this on the tree, we remove from Fig. 15.2 the alternative labelled "A" of this agent to the corresponding proposal.

Suppose, for example, that the type profile is (D, N_b, N_m), and that the second agent, a bilingual, uses sign language, which only D understands. Then the game tree is given by Fig. 15.3.[7]

15.3.2 The Unanimity Bargaining Games

A unanimity bargaining game requires that two responders have to take A to reach an agreement. The rest is exactly the same as in the majority bargaining game. Therefore, if everyone understands everyone else, then the game is given by Fig. 15.4.

---

[7]Although the tree corresponding to this case can be simplified, we leave it this way in order to clarify the rule for modifying the game when some agent does not understand the language used by the proposer.

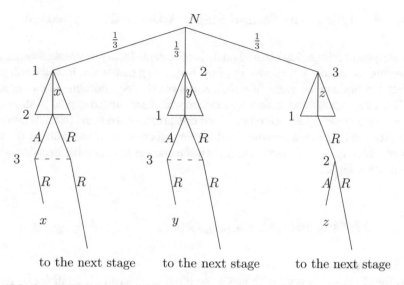

Fig. 15.3 A majority bargaining stage with (D, N_b, N_m) and N_b using sign language. This is the situation where 1 and 2 communicate with each other, 1 and 3 cannot communicate, and 2 understands 3, but 3 does not understand 2

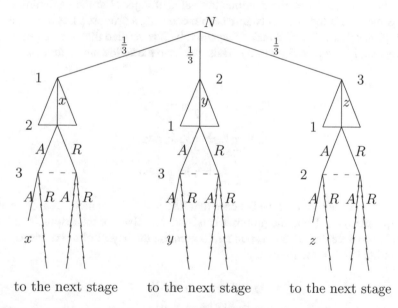

Fig. 15.4 A unanimity bargaining stage when everyone understands the language used in the bargaining

15.4   Analysis of the Second Stage: A Deductive Approach

The analysis of this section is based on deductive game theory where we assume that the agents are aware of the structure of the bargaining games and look for subgame perfect equilibria of the game. We often call it simply "equilibrium" in the sequel.

We look for a subgame perfect equilibrium with stationary strategies. A stationary strategy is the strategy according to which one proposes the same (mixed) outcome whenever she becomes a proposer, and for any outcome w, if one takes A for w at some decision node (information set), then she takes A to w at other decision nodes (information sets).

15.4.1   The Majority Bargaining Games

We divide the analysis into four cases.

The case of a single language This case occurs if the status profile is either (D, D, D) or (N_m, N_m, N_m). It is also verified that the profiles (D, D, N_b), (N_m, N_m, N_b), and (N_b, N_b, N_b) essentially correspond to this case.

Let V_i $(i = 1, 2, 3)$ be the continuation value of Agent i at the beginning of the stage game, i.e., at the node of Nature before choosing a proposer. Let x, y, and z be the proposals that 1, 2, and 3 make, respectively. It is verified that in an equilibrium, the proposal is accepted right away. Thus, we have the following equations.

$$V_1 = \frac{1}{3}x_1 + \frac{1}{3}y_1 + \frac{1}{3}z_1$$
$$V_2 = \frac{1}{3}x_2 + \frac{1}{3}y_2 + \frac{1}{3}z_2$$
$$V_3 = \frac{1}{3}x_3 + \frac{1}{3}y_3 + \frac{1}{3}z_3.$$

In an equilibrium, it must be the case that "A" and "R" are indifferent for one of the responders; for if not, the proposer can lower the share of the responder a little, which is still accepted by the responder, to increase the payoff of the proposer. This observation leads to the following.

$$x_2 = \delta V_2 \text{ or } x_3 = \delta V_3$$
$$y_1 = \delta V_1 \text{ or } y_3 = \delta V_3$$
$$z_1 = \delta V_1 \text{ or } z_2 = \delta V_2.$$

Once one expects to obtain "A" from one responder, this agent does not need to obtain "A" from the other. Thus, the proposer always offers zero to at least one of the responders.

We have three resource constraints (it is verified that the entire pie is divided among the agents).

$$x_1 + x_2 + x_3 = 1 \tag{15.1}$$
$$y_1 + y_2 + y_3 = 1 \tag{15.2}$$
$$z_1 + z_2 + z_3 = 1. \tag{15.3}$$

It is verified that $V_1 = V_2 = V_3$ holds. Let us check this assertion. Suppose first that $V_1 > V_2 > V_3$. Then Player 1 (resp. 2) tries to obtain "A" from 3 rather than 2 (resp. 1), and we have

$$x = (1 - \delta V_3, 0, \delta V_3)$$
$$y = (0, 1 - \delta V_3, \delta V_3)$$
$$z = (0, \delta V_2, 1 - \delta V_3).$$

This implies, for example,

$$V_1 = \frac{1}{3}(1 - \delta V_3) < \frac{1}{3}(1 - \delta V_3) + \frac{1}{3}\delta V_2,$$

which is a contradiction. Other cases with equalities involved are similarly shown to draw a contradiction.

Hence, we have the following proposition.

Proposition 15.1 *In the majority bargaining game, if the status profile is either one of* (D, D, D), (N_m, N_m, N_m), (D, D, N_b), (N_m, N_m, N_b), *and* (N_b, N_b, N_b), *then*

$$V_1 = V_2 = V_3 = \frac{1}{3}$$

holds for all $\delta \in (0, 1)$.

The case of no communication tool between majority and minority This game corresponds to the classical two-person bargaining game with random proposers. Consider (D, D, N_m). Since only 1 and 2 can communicate with each other, and their agreement stands, N_m is a dummy agent, obtaining nothing.

The continuation values of Agents 1 and 2 are given by

$$V_1 = \frac{1}{3}x_1 + \frac{1}{3}y_1 + \frac{1}{3}\delta V_1 \tag{15.4}$$
$$V_2 = \frac{1}{3}x_2 + \frac{1}{3}y_2 + \frac{1}{3}\delta V_2. \tag{15.5}$$

Solving this system with some other constraints, we obtain the following result.

Proposition 15.2 *In the majority bargaining game, if the status profile is either* (D, D, N_m) *or* (N_m, N_m, D), *then we have*

$$x = \left(\frac{3 - 2\delta}{3 - \delta}, \frac{\delta}{3 - \delta}, 0 \right),$$

$$y = \left(\frac{\delta}{3 - \delta}, \frac{3 - 2\delta}{3 - \delta}, 0 \right),$$

and

$$V = \left(\frac{1}{3 - \delta}, \frac{1}{3 - \delta}, 0 \right).$$

Moreover, as δ goes to one, x, y, and V all converge to $(1/2, 1/2, 0)$.

The case of two bilinguals This case occurs if the status profile is either (N_{b1}, N_{b2}, D) or (N_{b1}, N_{b2}, N_m). In this case, an important role is played by an additional substage prior to the bargaining game where two bilinguals simultaneously decide which language to use. Note that this substage is different from the first stage where the agents choose whether to become a bilingual or not.

Take (N_b, N_b, N_m) for the sake of argument. This game is more complicated than the previous case since the substage in which Agents 1 and 2 have moves to determine the language they use matters in a non-trivial way.

First of all, suppose that both Agents 1 and 2 choose m to accommodate 3. In this case, the outcome is the same as that of the standard case analyzed above, i.e.,

$$V_1 = V_2 = V_3 = \frac{1}{3}$$

holds for all $\delta \in (0, 1)$.

Next, suppose that one of the bilinguals, say, 1 chooses sign language, and if the other bilingual, 2, chooses oral language (or the both languages). In this situation, Agent 3 is still accommodated, and the above analysis applies. Agent 2 takes a balance between Agents 1 and 3 so that $V_1 = V_3$ holds. Moreover, since Agent 1 understands Agent 3's offer, Agent 3 takes a balance between 1 and 2 so that $V_1 = V_2$ holds. Thus, we have

$$V_1 = V_2 = V_3 = \frac{1}{3}$$

for all $\delta \in (0, 1)$.

If, on the other hand, two bilinguals both choose sign language, then Agent 3 (N_m) would never understand the two, and therefore, Agents 1 and 2 share the pie by themselves. In an equilibrium, it must be the case that A and R are indifferent for the responder as before. However, since neither Agent 1 nor 2 needs to worry about Agent 3's incentive, we have only two equations that correspond to this incentive constraint.

$$x_2 = z_2 = \delta V_2 \tag{15.6}$$
$$y_1 = z_1 = \delta V_1. \tag{15.7}$$

As for Agent 3, Agents 1 and 2 give nothing in their proposals. Therefore, we have

$$x_3 = y_3 = 0.$$

We have three resource constraints (15.1)–(15.3).

Solving this system of equations, we obtain

$$x = (x_1, x_2, x_3) = \left(\frac{3 - 2\delta}{3 - \delta}, \frac{\delta}{3 - \delta}, 0 \right), \tag{15.8}$$

$$y = (y_1, y_2, y_3) = \left(\frac{\delta}{3 - \delta}, \frac{3 - 2\delta}{3 - \delta}, 0 \right), \tag{15.9}$$

$$z = (z_1, z_2, z_3) = \left(\frac{\delta}{3 - \delta}, \frac{\delta}{3 - \delta}, 3\frac{1 - \delta}{3 - \delta} \right). \tag{15.10}$$

Thus, we have

$$V_1 = \frac{1}{3 - \delta},$$
$$V_2 = \frac{1}{3 - \delta},$$
$$V_3 = \frac{1 - \delta}{3 - \delta}.$$

Observe that $V_1 = V_2 > 1/3$, and that as δ goes to one, x, y, z, and V all converge to

$$\left(\frac{1}{2}, \frac{1}{2}, 0 \right).$$

Note that in the language choice substage prior to the bargaining game, choosing sign language is a weakly dominant strategy if we view this substage as a 2×2 game. We simply say that the players take a weakly dominant strategy in the substage of language selection, and we assume so in the sequel.

Proposition 15.3 *In the majority bargaining game, suppose that the status profile is either (N_b, N_b, D) or (N_b, N_b, N_m). Then, in a stationary subgame perfect equilibrium where the players take a weakly dominant strategy in the substage of language selection, the expected payoff profile is given by*

$$\left(\frac{1}{3 - \delta}, \frac{1}{3 - \delta}, \frac{1 - \delta}{3 - \delta} \right)$$

Moreover, as δ goes to one, the profile converges to $(1/2, 1/2, 0)$.

A deaf, a bilingual, and a monolingual This case occurs when we have (D, N_b, N_m). First, $V_1 = V_3 = V$ holds since if $V_1 > V_3$ (resp. $V_1 < V_3$) holds, then N_b favors 3 (resp. 1). Therefore, N_b uses the following strategy: N_b uses both sign and oral languages and offers $(\delta V, 1 - \delta V, 0)$ and $(0, 1 - \delta V, \delta V)$ with probability $1/2$ each.

Next, $x = (1 - \delta V_2, \delta V_2, 0)$ and $z = (0, \delta V_2, 1 - \delta V_2)$ hold since Agents 1 and 3 cannot communicate with each other.

Thus, we have

$$V = \frac{1}{3}(1 - \delta V_2) + \frac{1}{6}\delta V \tag{15.11}$$

$$V_2 = \frac{2}{3}\delta V_2 + \frac{1}{3}(1 - \delta V). \tag{15.12}$$

Solving this equation, we obtain the following result.

Proposition 15.4 *In the majority bargaining game, if the status profile is (D, N_b, N_m), then we have*

$$V_1 = V_3 = V = \frac{2}{3}\frac{(18 - \delta)(1 - \delta)}{(6 - \delta)(6 - 5\delta)}$$

$$V_2 = \frac{1}{3}\frac{4 - \delta}{6 - 5\delta}.$$

Moreover, as δ goes to one, V and V_2 converge to 0 and 1, respectively.

Summary of the majority bargaining game: the payoff table The payoff table in the limit of δ going to one is given by Table 15.1.

In this table, the entry that corresponds to row $i \in \{D, N_b, N_m\}$ and column $(j, k) \in \{D, N_b, N_m\}^2$ is the payoff of agent i who encounters (j, k).

15.4.2 The Unanimity Bargaining Games

Once the majority bargaining games are solved, the analysis of the unanimity bargaining games as set up in Sect. 15.3.2 is either reduced to some cases of the majority bargaining games or becomes trivial. The payoff table in the limit of δ going to one is given by Table 15.2. The table is read in the same manner as that for the majority bargaining game.

Table 15.1 Payoff table: majority bargaining games

	(D, D)	(D, N_b)	(N_b, N_b)	(D, N_m)	(N_b, N_m)	(N_m, N_m)
D	$\frac{1}{3}$	$\frac{1}{3}$	0	$\frac{1}{2}$	0	0
N_b	$\frac{1}{3}$	$\frac{1}{2}$	$\frac{1}{3}$	1	$\frac{1}{2}$	$\frac{1}{3}$
N_m	0	0	0	$\frac{1}{2}$	$\frac{1}{3}$	$\frac{1}{3}$

Table 15.2 Payoff table: unanimity bargaining games

	(D, D)	(D, N_b)	(N_b, N_b)	(D, N_m)	(N_b, N_m)	(N_m, N_m)
D	$\frac{1}{3}$	$\frac{1}{3}$	$\frac{1}{3}$	0	0	0
N_b	$\frac{1}{3}$	$\frac{1}{3}$	$\frac{1}{3}$	1	$\frac{1}{3}$	$\frac{1}{3}$
N_m	0	0	$\frac{1}{3}$	0	$\frac{1}{3}$	$\frac{1}{3}$

As we have pointed out earlier, the reason that we consider these two games is *not* because they are the only possible games played in the community. Rather, people in the community must have played a variety of games. In language selection games, however, it is often taken for granted that the more people learn a certain language, the more incentive one has to learn this language. We show that this claim needs to be examined. As we have seen in the case of the unanimity bargaining game, this is not the case. The game exhibits strategic substitutability, i.e., the more people learn a certain language, the less incentive one has to learn it. At the same time, however, it is shown that even if the unanimity bargaining game is played, people have more incentives to learn sign language on Martha's Vineyard Island than in the mainland USA. In other words, an incentive to become a bilingual is larger in the former than in the latter, but the bilinguals crowd out each other; given the population of the deaf, an incentive to become a bilingual is smaller when there are many bilinguals than when there are few.

One may casually argue that the more people use a certain language, the more incentive one has to learn it. This claim is not accurate at least in the present framework. We may state that the more people can use *only* a certain language, the more incentive one has to learn it. Again, we do not wish to make a general claim like this. The point is that the incentive to learn language is relative to the game they play, and that it does not contradict with the fact that even on the Island, there were people who did not use sign language.

15.5 The Analysis of the First Stage: An Evolutionary Approach

Using Tables 15.1 and 15.2, we examine the incentive of the non-deaf people to learn sign language. Suppose that the fraction of the deaf people is exogenously determined, and is denoted by α. Suppose next that the fraction of bilinguals in the non-deaf population is given by β. Then the probability that one meets, say, (D, N_b) is calculated as

$$2(1 - \alpha)\alpha\beta.$$

Table 15.3 shows the probabilities/frequencies of matching with respective pairs for all the cases.

Table 15.3 Frequency of matching

	(D, D)	(D, N_b)	(N_b, N_b)	(D, N_m)	(N_b, N_m)	(N_m, N_m)
Probability	α^2	$2(1-\alpha)\alpha\beta$	$(1-\alpha)^2\beta^2$	$2(1-\alpha)\alpha(1-\beta)$	$2(1-\alpha)^2$	$(1-\alpha)^2(1-\beta)^2$

15.5.1 The Majority Bargaining Game

Using Tables 15.1 and 15.3, we calculate the payoffs π_B^J and π_M^J of taking N_b and N_m, respectively, for the majority bargaining game.

$$\pi_B^J = \frac{1}{3}\alpha^2 + (1-\alpha)\alpha\beta + \frac{1}{3}(1-\alpha)^2\beta^2$$
$$+ 2(1-\alpha)\alpha(1-\beta) + (1-\alpha)^2 + \frac{1}{3}(1-\alpha)^2(1-\beta)^2\alpha^2$$
$$\pi_M^J = (1-\alpha)\alpha(1-\beta) + \frac{1}{3}(1-\alpha)^2 + \frac{1}{3}(1-\alpha)^2(1-\beta)^2\alpha^2.$$

Therefore, we have

$$\pi_B^J - \pi_M^J = \alpha\left(1 - \frac{2}{3}\alpha\right) + \frac{1}{3}(1-\alpha)^2\beta. \tag{15.13}$$

The coefficient of β is positive. Thus, the more bilinguals there are, the more incentive each agent has to learn sign language. The majority bargaining game exhibits strategic complementarity.

In order to analyze this situation, it is more appropriate to use evolutionary game theory than deductive game theory since the propagation of language, especially learning language in their childhood, is from parents to children and based on adaptation rather than deliberation. Let us use the best response dynamics introduced in Chap. 8 as an example.

Suppose that the initial condition is $\beta = 0$. A non-deaf agent prefers bilingual to monolingual at $\beta = 0$ if

$$d < d^* \equiv \alpha\left(1 - \frac{2}{3}\alpha\right). \tag{15.14}$$

Due to strategic complementarity, once the first non-deaf agent begins to learn sign language, all other agents start learning it, i.e., this process continues until every agent learns sign language (see Fig. 15.5).

If we use the historical data from the case of Martha's Vineyard, the threshold d^* on the island was about 20 times higher than that in mainland USA, and d^* in Chilmark was about 240 times higher than that on the mainland. This wide margin might make persons without disability adjust to the hereditary deafness on the island.

Fig. 15.5 Majority case

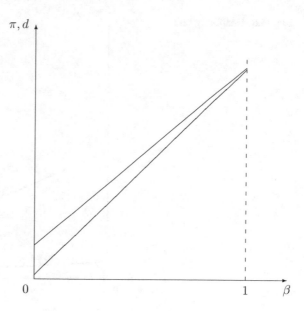

15.5.2   The Unanimity Bargaining Game

It is commonly assumed that language acquisition exhibits strategic complementarity, but this depends on the context. In the case of the unanimity bargaining game, we have strategic substitute, and its analysis becomes different from the case of strategic complement.

Using Tables 15.2 and 15.3, we calculate the payoffs of taking N_b and N_m, respectively, for the unanimity bargaining game.

$$\pi_B^U = \frac{1}{3} + \frac{4}{3}\alpha(1-\alpha)(1-\beta)$$

$$\pi_M^U = \frac{1}{3}(1-\alpha)^2.$$

Therefore, we have

$$\pi_B^U - \pi_M^U = \frac{1}{3}\left[\alpha(6-5\alpha) - 4(1-\alpha)\alpha\beta\right]. \tag{15.15}$$

The coefficient of β is negative. Thus, the more bilinguals there are, the less incentive each agent has to learn sign language (see Fig. 15.6). The unanimity bargaining game exhibits strategic substitute.

Fig. 15.6 Unanimity case

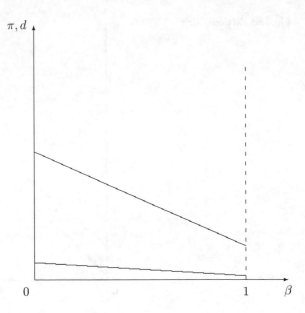

In this case, the threshold d^{**} to make the first agent have an incentive to learn sign language is given by

$$d^{**} = \frac{1}{3}\alpha(6 - 5\alpha).$$

As in the case of majority bargaining games, d^{**} on the island was about 40 times higher than that on the US mainland, and d^{**} in Chilmark was about 240 times higher. Unlike the previous case, however, since the unanimity bargaining game exhibits strategic substitute, d^{**} alone cannot explain the dynamics of people's choices.

Suppose that $d < d^{**}$ holds. Then persons without disability start learning sign language, but it stops at

$$\bar{\beta} \equiv \frac{\alpha(6 - 5\alpha) - 3d}{4\alpha(1 - \alpha)}$$

if $d > \frac{1}{3}\alpha(2 - \alpha)$; otherwise, all of the persons without disability become bilinguals.

15.6   Prejudice: An Inductive Approach

In the 1970s and 80s, there were a series of attempts to refine Nash equilibrium, the central solution concept in noncooperative game theory, from the viewpoint of the rationality hypothesis. This so-called refinement program reached its highest when Kohlberg and Mertens (1986) proposed the concept of socially stable sets. It is often assumed that players have sufficient knowledge of the structure of the game they play (at least in a probabilistic manner), and that they deduce what they should do (strategies) through reasoning processes. Here, let us call this type of game theory *deductive game theory*.

Its powerful hypothesis has not been well challenged. Recent studies on bounded rationality, or behavioral economics, only attest to it. In order to explain the phenomena that seem to deviate from the rational behavior, one needs to modify the rule of the game and/or preferences of the players therein. However, this modification itself reinforces the rationality hypothesis of game theory.

This rationality hypothesis, including the hypothesis behind recently developed behavioral economics, limits the scope of application of game theory to some issues associated with disability studies, which have been concerned with concepts like prejudices and social inclusion. Taking the concept of prejudice as an example, the Cambridge Dictionary online defines prejudice as "an unfair and unreasonable opinion or feeling, especially when formed without enough thought or knowledge." This concept, by nature, involves some sort of wrong beliefs and limited reasoning. If it is assumed that players know the structure of the world, then it leads to the world with prejudices. On the other hand, if one constructs a model in which players have wrong beliefs, such a model presumes prejudices as an outset, and we cannot study how and in what circumstances prejudices emerge.

To understand societal phenomena like the emergence of prejudices without assuming it at the outset, we consider a situation where players do not know the entire structure of the game, obtain information through the course of the playing the game, and try to construct a model of the game. A *model* of the game itself is described as a game, which may or may not be the same as the original game. A model is *coherent* with a set of information (a priori knowledge and experiences) if it does not contradict with prior knowledge, if any, and the information acquired through the play of the game (experiences). We apply this concept to the game we have analyzed so far and study what type of model people might construct through playing the game.

Also, when we discuss inclusive education, one of its purposes is to enlighten students without disability that our society consists of a variety of people through the interaction between persons with and without disability. To tell a story like this, we are faced with the limit of the rationality hypothesis since rational players are aware of this possibility, and their knowledge acquisition is by eliminating some possible states of the world out of scope rather than by adding new insight from their experiences and information they obtain.

In order to construct a theory of economy and disability, incorporating some ideas that have been discussed in the disability community, we introduce a new theory, inductive game theory, which was first developed by Kaneko and Matsui (1999) (see Chaps. 13 and 14). In this theory, players do not know the structure of the game they play. Instead, they accumulate experiences and try to understand the society, thereby constructing a model of the society. The constructed model may or may not correspond to the "true" model, if indeed the true model ever exists. This way, inductive game theory is able to describe the situation where people, based on their limited experiences, form a false model of the world. This false image may include prejudice as a typical example.

For this purpose, let us consider a specific situation. Suppose first that nobody without disability learns sign language, playing unanimity bargaining games. It may not occur to them that there is a choice of learning it.

Suppose next that each player in N_m sees the structure of the subgame of the bargaining stage, including the payoff therein. However, they do not see its structure unless they enter the subgame that they actually play. For example, suppose the player in question is in N_m. Then it is only after this player meets a player in D and a player in N_m to enter the subgame where (N_m, N_m, D) is a type profile that this player sees the payoff structure of this subgame.

Suppose now that the player plays the unanimity bargaining game several times. The experience she obtains is the following:

- met with (D, D), and no agreement was reached;
- met with (D, N), and no agreement was reached;
- met with (N, N), and an agreement was reached, and each member obtains $1/3$.

Note that as experiences, the player thinks the non-deaf people who do no use sign language are N instead of N_m.

One of the simplest payoff function that is coherent with the above knowledge and experiences is given by

$$U_N(i, j) = \begin{cases} \frac{1}{3} & \text{if } i, j = N, \\ 0 & \text{otherwise,} \end{cases}$$

where $U_N(i, j)$ is the subgame perfect equilibrium payoff when this player meets a player of type i and a player of type j. Then the situation in their eyes can be summarized as in Table 15.4. Notice that there is no distinction between N_b and N_m, while in reality, everyone is in N_m.

Of course, this is a simplistic model compared to the true model laid out in the present analysis. However, its explanatory power is as good as the true model given the experiences they have. Moreover, this model may be regarded as a better model than the true model precisely because it describes the world in simpler terms than the true one. Prejudices emerge.

This suggests, for example, the importance of inclusive education, i.e., both children with and without disability learn at the same school, side by side.

In Japan, children with disability are often go to special support schools. There are pros and cons in this system. For children with hearing impairment, they are better educated by sign language. Groce (1985) indicates that deaf children using

Table 15.4 Payoff table: unanimity bargaining games from non-deaf's viewpoint

	$\{D, D\}$	$\{D, N\}$	$\{N, N\}$
D	0	0	0
N	0	0	$\frac{1}{3}$

sign language quickly acquire a vocabulary of about 1000 words by the age of five, which is almost as many as hearing children. However, if they are taught by oral instruction only, then they learn only several dozen words.

In spite of findings of this type, to reduce prejudices of children without hearing impairment, it is important to include deaf children in regular schools or nurseries where teachers or childcare workers can use both oral and sign languages. If one sees there is a bilingual state of this kind, the students can see that among those who can communicate, there are both bilingual and monolingual people, i.e., N_b and N_m, and they realize that the payoff table like that in Table 15.4 is a false image of the world.

15.7 Remarks

The present analysis takes Martha's Vineyard Island as an example to show that disability is a concept that is relative to the society. Similar analyses, yet different in details, are applicable to a variety of issues that face our society. One thing in common is that disability is a social construct.

The present analysis is suggestive rather than definitive, conveying the idea that disability-related issues are societal phenomena, emerging from the interactions of people therein, and that game theory provides us with a useful approach to these issues.

Disability-related issues call for a new approach in game theory, especially when we discuss the issues of prejudices and inclusion. Inductive game theory is one of such examples.

Chapter 16
Concluding Remarks

We have presented a game-theoretic analysis of economy and disability, although the analysis is not confined to the disability-related issues. We have not conducted traditional economic analysis such as cost-benefit analysis. Rather, we have focused on human interaction and tried to account for some disability-related phenomena. We have emphasized the majority-minority aspects of the issues as opposed to the difference between persons *with* disability and persons *without*. We have first discussed various cases that motivate the present study. After introducing evolutionary dynamics, we have examined the interaction of multiple groups with different traits and/or preferences. We have also analyzed the mechanism under which prejudices, negative preferences, emerge as a result of segregation. This is a new approach in game theory since economic and game theories have presumed preferences of the agents/players as given and studied their behavior. This analysis has shown the potential benefit of inclusive education as opposed to separate education. The last chapter before conclusion was about the case of hereditary deafness in Martha's Vineyard Island in the past centuries. It has provided us with a game-theoretic account of the inclusion of people using sign language: the solution was that most people on the island, irrespective of deafness, used sign language as a matter of course. Consequently, the deafness was not considered as disability there.

Significant part of the present book is devoted to the analysis of formal models. The reader might think why we need such mathematical and abstract models, which should be well taken. I have a word on this, however. When I was a graduate student at Northwestern University, Professor Don Saari taught us the first year graduate mathematics. He was a professor in mathematics, but he also wrote excellent papers in theoretical economics. One day, he received a question from a female student. She asked why we need to study such an abstract math instead of an elementary math, which is more intuitive and imaginable than the abstract one. Professor Saari smiled and asked us a question, "What is the part of speech of the term 'abstract'?" He continued to say that we might think it is an adjective, but we should regard it as a *verb*. He said that we abstracted the essence of mathematics with a gesture

© Springer Nature Singapore Pte Ltd. 2019
A. Matsui, *Economy and Disability*, Economy and Social Inclusion,
https://doi.org/10.1007/978-981-13-7623-8_16

of grabbing and taking out something in the air. If you look into thesaurus,[1] the synonyms of "abstract" as an adjective are "abstruse", "unreal", "hypothetical", and so on. On the other hand, the synonyms of the same word as a verb are "trim", "summarize" and "digest", among others. It was an enlightening moment for me. A theoretical account often abstracts the reality. However, it does not make the story unreal as some people might think, but it trims details, summarizes the reality and digests it.

Let us consider a simple example. Prisoners' dilemma is one of the most famous examples in game theory. Two prisoners' who committed a crime are questioned by a prosecutor who need an evidence. Without confession, it is difficult to obtain the evidence. The prosecutor separates the prisoners and tells each of them independently that if he confesses, then a special treatment awaits him, but if the other prisoner confesses first, then he bears all the responsibility of the crime. In this situation, remaining silent is a risky strategy as the other person has an incentive to confess. The prisoners end up with confession, which leads to an inefficient outcome for them.

If prisoners' dilemma could explain the prisoners' dilemma only, it would not have been this famous. The game has become famous as the same game captures the essence of price competition in economics and that of armament race in politics. In the case of price competition, the players are firms, and "remain silent" should read "high price", while "confess" should read "low price". A firm is hit hard if it chooses "high price" and if the opponent chooses "low price". Thus, although it is beneficial for the firms to cooperate by taking "high price" together, they cannot cooperate and end up with competing with each other.

In the case of armament race, the players are two nations, and "remain silent" should read "disarmament", while "confess" should read "armament". Here, "disarmament" is risky since the opponent has an incentive to choose "armament" and if that is chosen, the nation's decision "disarmament" will put the nation in a vulnerable position. As a result, the two nations fail to choose "disarmament". Treating these three situations in the same manner would be impossible if we did not abstract the phenomena.

In the realm of disability, the theory gives us a strong account of disability being a social construct. First, consider our society where we have stairs in many places so that persons with wheelchair cannot go to as many places as they wish to. Second, consider another society, which is hypothetical, where most people are like athletes in gymnastics so that a climbing pole suffices in order to go up to the upper floors. Stairs are so expensive to build that this society would put climbing poles everywhere, but not stairs since few people need stairs anyway. In such a society, we, the "normal" persons in our world, become persons with disability since we cannot go up to the second floor.

If we build a simple yet formal model to explain the first society, it would also explain the second hypothetical society. In the second society, we, who cannot climb the pole, are called persons with disability.

---

[1]See   http://www.thesaurus.com/browse/abstract   and   https://www.merriam-webster.com/ thesaurus/abstract. Last check: June 7, 2018.

All the formal models, like all the informal arguments, have lots of flaws. The fate of models is that they simplify the real world. There are many elements of the world that models fail to take into account. However, as Joan Robinson put it,

A map at the scale of 1:1 is of no use to a traveller. —Joan Robinson[2]

Models are useful in understanding the nature of the world just like a map is useful in going to a destination. A nice thing about formal models is that missing pieces of the models are universally recognizable. If the reader finds the missing pieces in the models presented in this book, incorporates them into her model, and advances the understanding of economy and disability, the goal of this book will be achieved.

---

[2]See p. 174, Harcourt and Kerr (2009).

References

Akerlof, G. A., & Kranton, R. E. (2000). Economics and identity. *The Quarterly Journal of Economics, 115*(3), 715–753.

Aoki, M. (1994). The Japanese firm as a system of attributes: A survey and research agenda. *The Japanese Firm: Sources of Competitive Strength* (pp. 11–40). New York: Doubleday.

Aoki, M. (2001). *Toward a comparative institutional analysis*. Cambridge, MA: MIT Press.

Asimov, I. (1995). *Yours, Isaac Asimov: A life in letters*.

Aubin, J. P., & Cellina, A. (1984). *Differential inclusions: Set-valued maps and viability theory: With 29 figures*. Berlin: Springer.

Aumann, R. J. (1987). Correlated equilibrium as an expression of Bayesian rationality. *Econometrica: Journal of the Econometric Society*, 1–18.

Aumann, R. J., & Sorin, S. (1989). Cooperation and bounded recall. *Games and Economic Behavior, 1*(1), 5–39.

Battigalli, P., & Siniscalchi, M. (2002). Strong belief and forward induction reasoning. *Journal of Economic Theory, 106*(2), 356–391.

Becker, G. S. (1957). *The economics of discrimination*. University of Chicago.

Bernheim, B. D. (1984). Rationalizable strategic behavior. *Econometrica: Journal of the Econometric Society*, 1007–1028.

Bernheim, B. D., Peleg, B., & Whinston, M. D. (1987). Coalition-proof Nash equilibria I. *Concepts. Journal of Economic Theory, 42*(1), 1–12.

Binmore, K. (October, 1987). Modeling rational players: Part I. *Economics and Philosophy, 3*(2), 179–214. ISSN 0266-2671.

Binmore, K. (1988). Modeling rational players: Part II. *Economics & Philosophy, 4*(1), 9–55.

Binmore, K. G., & Samuelson, L. (1992). Evolutionary stability in repeated games played by finite automata. *Journal of Economic Theory, 57*(2), 278–305.

Brandenburger, A., & Dekel, E. (1987). Rationalizability and correlated equilibria. *Econometrica: Journal of the Econometric Society, 55*, 1391–1402.

Brown, G. W. (1951). Iterative solution of games by fictitious play. *Activity Analysis of Production and Allocation, 13*(1), 374–376.

Cabinet Office of Japan. (2017a). Shogaisha sesaku kanren yosan-no gaiyo (outline of the budget for disability-related policies). http://www8.cao.go.jp/shougai/suishin/yosan/pdf/yosan.pdf. Last view: June 22, 2018.

© Springer Nature Singapore Pte Ltd. 2019
A. Matsui, *Economy and Disability*, Economy and Social Inclusion,
https://doi.org/10.1007/978-981-13-7623-8

Cabinet Office of Japan. (2017b). Shogaisha hakusho (annual report on government measures for persons with disabilities). http://www8.cao.go.jp/shougai/whitepaper/h29hakusho/zenbun/index-pdf.html. Last view: June 22, 2018.

Camerer, C. F. (2003). *Behavioral game theory: Experiments in strategic interaction*. Princeton: Princeton University Press.

Cavalli-Sforza, L. L., & Feldman, M. W. (1981). *Cultural transmission and evolution: A quantitative approach*. Princeton: Princeton University Press.

Coddington, E. A., & Levinson, N. (1955). *Theory of ordinary differential equations*. New York: Tata McGraw-Hill Education.

Crawford, V. P. (1991). An "evolutionary" interpretation of Van Huyck, Battalio, and Beil's experimental results on coordination. *Games and Economic Behavior, 3*(1), 25–59.

Crawford, V. P., & Sobel, J. (1982). Strategic information transmission. *Econometrica: Journal of the Econometric Society, 50*, 1431–1451.

Desforges, J. F., Wintrobe, M. M., Furie, B., & Schwartz, R. S. (2018). Blood disease. Britannica.com. https://www.britannica.com/science/blood-disease. Last view: June 23, 2018.

Ellison, G. (1993). Learning, local interaction, and coordination. *Econometrica: Journal of the Econometric Society, 61*, 1047–1071.

Elster, J. (1989). *The cement of society: A survey of social order*. Cambridge: Cambridge University Press.

Encyclopaedia Britannica. (2011). Yogachara. Britannica.com. https://www.britannica.com/topic/Yogachara. Last view: June 23, 2018.

Farrell, J. (1993). Meaning and credibility in cheap-talk games. *Games and Economic Behavior, 5*(4), 514–531.

Foster, D., & Young, P. (1990). Stochastic evolutionary game dynamics. *Theoretical Population Biology, 38*(2), 219–232.

Fudenberg, D., & Harris, C. (1992). Evolutionary dynamics with aggregate shocks. *Journal of Economic Theory, 57*(2), 420–441.

Fudenberg, D., & Levine, D. K. (1993). Self-confirming equilibrium. *Econometrica: Journal of the Econometric Society, 61*, 523–545.

Fudenberg, D., & Levine, D. K. (1998). *The theory of learning in games*. Cambridge: MIT Press.

Fudenberg, D., & Tirole, J. (1991). *Game theory*. Cambridge, MA: MIT Press.

Fundenberg, D., & Maskin, E. (1990). Evolution and cooperation in noisy repeated games. *The American Economic Review, 80*(2), 274–279.

Gilboa, I., & Matsui, A. (1991). Social stability and equilibrium. *Econometrica: Journal of the Econometric Society, 59*, 859–867.

Gilboa, I., & Matsui, A. (1992). A model of random matching. *Journal of Mathematical Economics, 21*(2), 185–197.

Gilboa, I., & Schmeidler, D. (1995). Case-based decision theory. *The Quarterly Journal of Economics, 110*(3), 605–639.

Gilboa, I., & Schmeidler, D. (2001). *A Theory of Case-Based Decisions*. Cambridge University Press.

Groce, N. E. (1985). *Everyone here spoke sign language*. Cambridge: Harvard University Press.

Hammerstein, P., & Selten, R. (1994). Game theory and evolutionary biology. *Handbook of Game Theory with Economic Applications, 2*, 929–993.

Harcourt, G., & Kerr, P. (2009). *Joan Robinson*. Berlin: Springer.

Hart, O. D. (1975). On the optimality of equilibrium when the market structure is incomplete. *Journal of Economic Theory, 11*(3), 418–443.

Hendron, J. (2000). Return to the wild. *Endangered Species, 25*(3), 10–11.

Hofbauer, J., & Sandholm, W. H. (2002). On the global convergence of stochastic fictitious play. *Econometrica, 70*(6), 2265–2294.

Hofbauer, J., & Sigmund, K. (1998). *Evolutionary games and population dynamics*. Cambridge, MA: Cambridge University Press.

Hume, D. (1739/1984). *A treatise of human nature*. Penguin Classics.

Kalai, E., & Lehrer, E. (1995). Subjective games and equilibria. *Games and Economic Behavior*, *8*(1), 123–163. ISSN 0899-8256.

Kalai, E., & Samet, D. (1984). Persistent equilibria in strategic games. *International Journal of Game Theory*, *13*(3), 129–144.

Kandori, M. (1997). Evolutionary game theory in economics. *Econometric Society Monographs*, *26*, 243–277.

Kandori, M., Mailath, G. J., & Rob, R. (1993). Learning, mutation, and long run equilibria in games. *Econometrica: Journal of the Econometric Society*, *61*, 29–56.

Kandori, M., & Rob, R. (1995). Evolution of equilibria in the long run: A general theory and applications. *Journal of Economic Theory*, *65*(2), 383–414.

Kaneko, M. (1987). The conventionally stable sets in noncooperative games with limited observations I: Definitions and introductory arguments. *Mathematical Social Sciences*, *13*(2), 93–128 (ISSN 0165-4896).

Kaneko, M. (1998). *Evolution of thoughts: Deductive game theories in the inductive game situation*. Center for Economic Research, Tilburg University.

Kaneko, M., & Kimura, T. (1992). Conventions, social prejudices and discrimination: A festival game with merrymakers. *Games and Economic Behavior*, *4*(4), 511–527.

Kaneko, M., & Kline, J. J. (2008). Inductive game theory: A basic scenario. *Journal of Mathematical Economics*, *44*(12), 1332–1363.

Kaneko, M., & Matsui, A. (1999). Inductive game theory: Discrimination and prejudices. *Journal of Public Economic Theory*, *1*(1), 101–137.

Katz, K., & Matsui, A. (2004). When trade requires coordination. *Journal of the Japanese and International Economies*, *18*(3), 440–461.

Kawagoe, T., & Matsui, A. (2012). Economics, game theory and disability studies. In A. Azzopardi & S. Grech (Eds.), *Inclusive communities* (pp. 119–131). Sense Publishers.

Kawashima, S. (2011). The term 'disability' in discrimination law. In A. Matsui, O. Nagase, A. Sheldon, D. Goodley, Y. Sawada, & S. Kawashima (Eds.), *Creating a society for all: Disability and economy* (pp. 107–116). Leeds: Disability Press.

Kodama, Y., Morozumi, R., Matsumura, T., Kishi, Y., Murashige, N., Tanaka, Y., et al. (2012). Increased financial burden among patients with chronic myelogenous leukaemia receiving imatinib in Japan: A retrospective survey. *BMC Cancer*, *12*(1), 152.

Kohlberg, E., & Mertens, J.-F. (1986). On the strategic stability of equilibria. *Econometrica: Journal of the Econometric Society*, *54*, 1003–1037.

Kreps, D. M., & Wilson, R. (1982). Sequential equilibria. *Econometrica: Journal of the Econometric Society*, *50*, 863–894.

Krugman, P. (1991). Increasing returns and economic geography. *Journal of Political Economy*, *99*(3), 483–499.

Lazear, E. P. (1999). Culture and language. *Journal of Political Economy*, *107*(6), 95–126.

Mailath, G. J. (1992). Introduction: Symposium on evolutionary game theory. *Journal of Economic Theory*, *57*, 259–277.

Mailath, G. J. (1993). Perpetual randomness in evolutionary economics. *Economics Letters*, *42*(2–3), 291–299.

Mailath, G. J. (1995). *Recent development in evolutionary game theory*. New York: Mimeo.

Marger, M. N. (1991). *Race and ethnic relations*, 2nd edn. Belmont: Wadsworth Publishing.

Mas-Colell, A., Whinston, M. D., Green, J. R., et al. (1995). *Microeconomic theory* (Vol. 1). New York: Oxford University Press.

Matsui, A. (1989). Information leakage forces cooperation. *Games and Economic Behavior*, *1*(1), 94–115.

Matsui, A. (1991). Cheap-talk and cooperation in a society. *Journal of Economic Theory*, *54*(2), 245–258.

Matsui, A. (1992). Best response dynamics and socially stable strategies. *Journal of Economic Theory*, *57*(2), 343–362.

Matsui, A. (2008). A theory of man as a creator of the world. *The Japanese Economic Review*, *59*(1), 19–32.

Matsui, A. (2017). Disability and economy: A game theoretic approach. *The Japanese Economic Review*, *68*(1), 5–23.

Matsui, A., Morozumi, R., Kaneko, Y., Kano, K., Kawamura, M., Sawada, Y., et al. (2012). Shogaisha-no nichijo/keizai katsudo chosa (survey on daily and economic activities of persons with disability). http://www.rease.e.u-tokyo.ac.jp/read/jp/archive/statistics/dantai_main. html. Last view: June 22, 2018.

Matsui, A., & Matsuyama, K. (1995). An approach to equilibrium selection. *Journal of Economic Theory*, *65*(2), 415–434.

Matsui, A., & Okuno-Fujiwara, M. (2002). Evolution and the interaction of conventions. *Japanese Economic Review*, *53*, 141–153.

Matsui, A., & Shimizu, T. (2007). *Abductive inference in game theory*. Technical report, Mimeo, University of Tokyo.

Matsuyama, K. (1991). Increasing returns, industrialization, and indeterminacy of equilibrium. *The Quarterly Journal of Economics*, *106*(2), 617–650.

Matsuyama, K., Kiyotaki, N., & Matsui, A. (1993). Toward a theory of international currency. *The Review of Economic Studies*, *60*(2), 283–307.

Maynard Smith, J., & Price, G. R. (1973). The logic of animal conflict. *Nature*, *246*(5427), 15–18.

Merton, R. K. (1949). Discrimination and the American creed. In *Discrimination and national welfare* (pp. 99–126).

Ministry of Education Culture Sports Science and Technology. (2015). Team-toshiteno gakko-no arikata-to konngo-no kaizen-housaku-nitsuite (the current and the future states of schools as a team): Reference materials. http://www.mext.go.jp/b_menu/shingi/chukyo/chukyo0/gijiroku/__icsFiles/afieldfile/2015/12/28/1365606_3_4.pdf. Last view: June 22, 2018.

Ministry of Health, Labour and Welfare. (2017). Shogai hoken fukushi-bu yosan-an-no gaiyo (outline of the budget of the division of disability, health and welfare). http://www.mhlw.go.jp/file/05-Shingikai-12601000-Seisakutoukatsukan-Sanjikanshitsu_Shakaihoshoutantou/0000147371.pdf. last view: June 22, 2018.

Miyasawa, K. (1961). *On the convergence of the learning process in a* 2×2 *non-zero-sum two person game*. Technical Report 33, Princeton University.

Monderer, D., & Shapley, L. S. (1996). Fictitious play property for games with identical interests. *Journal of Economic Theory*, *68*(1), 258–265.

Myerson, R. B. (1978). Refinements of the nash equilibrium concept. *International Journal of Game Theory*, *7*(2), 73–80.

Nash, J. (1951). Non-cooperative games. *Annals of Mathematics*, *54*, 286–295.

Nassau, K. (2018). Colour. Britannica.com. https://www.britannica.com/science/color. Last view: June 23, 2018.

OECD. (2003). *Transforming disability into ability: Policies to promote work and income security for disabled people*. Organisation for Economic Co-operation and Development.

OECD. (2013). OECD data. https://data.oecd.org/socialexp/social-spending.htm. Last view: June 22, 2018.

Okada, A. (1996). A noncooperative coalitional bargaining game with random proposers. *Games and Economic Behavior*, *16*(1), 97–108.

Okuno-Fujiwara, M. (1988). Interdependence of industries, coordination failure and strategic promotion of an industry. *Journal of International Economics*, *25*(1–2), 25–43.

Oliver, M. (1984). The politics of disability. *Critical Social Policy*, *4*(11), 21–32.

Oliver, M. (1996). *Understanding disability: From theory to practice*. New York: St. Martin's Press.

Oliver, M. (2013). The social model of disability: Thirty years on. *Disability & Society*, *28*(7), 1024–1026.

Pearce, D. G. (1984). Rationalizable strategic behavior and the problem of perfection. *Econometrica Journal of the Econometric Society*, *52*, 1029–1050.

Peirce, C. S. (1898/1992). *Reasoning and the logic of things: The Cambridge conferences lectures of 1898*. Harvard University Press.

Pepperberg, I. M. (1994). Numerical competence in an African gray parrot (Psittacus erithacus). *Journal of Comparative Psychology, 108*(1), 36–44.

Plato. (1941). *The republic of Plato* (trans: Cornford, F. M.). Oxford University Press.

Rabin, M. (1990). Communication between rational agents. *Journal of Economic Theory, 51*(1), 144–170.

Robson, A. J. (1990). Efficiency in evolutionary games: Darwin, nash and the secret handshake. *Journal of Theoretical Biology, 144*(3), 379–396. ISSN 0022-5193.

Sacks, O. (1996). *The Island of the Colour-blind and Cycad Island*. Picador.

Sandholm, W. H. (2010). *Population games and evolutionary dynamics*. Cambridge, MA: MIT Press.

Schilpp, P. A. (Ed.) (1979). *Albert Einstein: autobiographical notes*. Open Court.

Sclten, R. (1975), Reexamination of the perfectness concept for equilibrium points in extensive games. *International Journal of Game Theory, 4*(1), 25–55.

Shapley, L. S. (1962). On the nonconvergence of fictitious play. The RAND Corporation, RM-3026.

Shapley, L. S. (1964). Some topics in two-person games. *Advances in Game Theory, 52*, 1–29.

Shimono, K. (2000). Soutaiteki-kiken-kaihido no sokutei (estimation of relative risk aversion). *Oikonomika, 37*(1), 1–14.

Simon, H. A. (1957). *Models of man; social and rational*. Hoboken: Wiley.

Smith, H. L. (1988). Systems of ordinary differential equations which generate an order preserving flow. A survey of results. *SIAM Review, 30*(1), 87–113.

Smith, A. (2010). *The theory of moral sentiments*. London: Penguin.

Steiner, W. G. (2016). Drug use. Britannica.com. https://www.britannica.com/topic/drug-use. Last view: June 23, 2018.

Sugino, A. (2007). *Shogaigaku: Riron-Keisei to Shatei (Disability Studies: Theoretical Creation and Scope*. University of Tokyo Press.

Swinkels, J. M. (1992). Evolutionary stability with equilibrium entrants. *Journal of Economic Theory, 57*(2), 306–332.

Taylor, P. D., & Jonker, L. B. (1978). Evolutionary stable strategies and game dynamics. *Mathematical Biosciences, 40*(1–2), 145–156.

Thomas, B. (1985). On evolutionarily stable sets. *Journal of Mathematical Biology, 22*(1), 105–115.

Thorndike, E. (2017). *Animal intelligence: Experimental studies*. Abingdon: Routledge.

United Nations. (January 2019). Monthly bulletin of statistics.

van Damme, E. (1987). *Stability and perfection of Nash equilibria*. Berlin, New York: Springer. ISBN 0387171010.

van Damme, E. (1991). *Stability and perfection of Nash equilibria* (Vol. 339). Berlin: Springer.

van Damme, E. (1994). Evolutionary game theory. *European Economic Review, 38*(3), 847–858. ISSN 0014-2921.

Vega-Redondo, F. (1996). *Evolution, games, and economic behaviour*. Oxford: Oxford University Press.

von Neumann, J., & Morgenstern, O. (1947/2007). *Theory of games and economic behavior (commemorative edition)*. Princeton: Princeton University Press.

Weibull, J. W. (1997). *Evolutionary game theory*. Cambridge, MA: MIT Press.

Yamamori, T., Kato, K., Kawagoe, T., & Matsui, A. (2008). Voice matters in a dictator game. *Experimental Economics, 11*(4), 336–343.

Yamamori, T., Kato, K., & Matsui, A. (2010). When you ask zeus a favor: The third party's voice in a dictator game. *The Japanese Economic Review, 61*(2), 145–158.

Yoshino, Y. (2016). *3600-nichi-no Kiseki: Gan-to Tatakau Maihime (The miracle of 3600 days: the dancer fighting with cancer)*. Seisa University Press.

Young, H. P. (1993). The evolution of conventions. *Econometrica: Journal of the Econometric Society, 61*, 57–84.

Zeeman, E. C. (1981). Dynamics of the evolution of animal conflicts. *Journal of Theoretical Biology, 89*(2), 249–270.

Author Index

A

Akerlof and Kranton (2000), 126
Aoki (1994), 51
Aoki (2001), 65
Asimov (1995), 124
Aubin and Cellina (1984), 103
Aumann (1987), 55, 69
Aumann and Sorin (1989), 98

B

Battigalli and Siniscalchi (2002), 147
Becker (1957), 147, 194
Bernheim (1984), 96
Bernheim et al. (1987), 95
Binmore (1987), 55, 56, 69
Binmore (1988), 55, 56, 69
Binmore and Samuelson (1992), 79
Brandenburger and Dekel (1987), 69
Brown (1951), 74

C

Cabinet Office of Japan (2017a), 5
Cabinet Office of Japan (2017b), 4
Camerer (2003), 146
Cavalli-Sforza and Feldman (1981), 142
Coddington and Levinson (1955), 76
Crawford (1991), 52
Crawford and Sobel (1982), 106

D

Desforges et al. (2018), 27

E

Ellison (1993), 120

E

Elster (1989), 52, 53, 120
Encyclopaedia Britannica (2011), 158

F

Farrell (1993), 105
Foster and Young (1990), 91, 120
Fudenberg and Harris (1992), 91
Fudenberg and Levine (1993), 147, 162
Fudenberg and Levine (1998), 74
Fundenberg and Maskin (1990), 79
Fudenberg and Tirole (1991), 169

G

Gilboa and Matsui (1991), 60, 69, 74, 87
Gilboa and Matsui (1992), 93
Gilboa and Schmeidler (1995), 147, 162
Groce (1985), 25, 202, 216

H

Hammerstein and Selten (1994), 52
Harcourt and Kerr (2009), 221
Hart (1975), 126
Hendron (2000), 145
Hofbauer and Sandholm (2002), 74
Hofbauer and Sigmund (1998), 52, 60
Hume (1739/1984), 154, 155

K

Kalai and Lehrer (1995), 162
Kalai and Samet (1984), 69, 71
Kandori (1997), 201
Kandori and Rob (1995), 92
Kandori et al. (1993), 91, 120

© Springer Nature Singapore Pte Ltd. 2019
A. Matsui, *Economy and Disability*, Economy and Social Inclusion,
https://doi.org/10.1007/978-981-13-7623-8

Kaneko (1987), 55, 69
Kaneko (1998), 194
Kaneko and Kimura (1992), 160
Kaneko and Kline, 147
Kaneko and Matsui (1999), 142, 147, 159,
    174, 215
Katz and Matsui (2004), 123
Kawagoe and Matsui (2012), 64
Kawashima (2011), 14
Kodama et al. (2012), 27, 28
Kohlberg and Mertens (1986), 69, 71, 89,
    214
Kreps and Wilson (1982), 95
Krugman (1991), 121

L
Lazear (1999), 125

M
Mailath (1992), 52
Mailath (1993), 52
Mailath (1995), 52
Marger (1991), 193
Mas-Colell et al. (1995), 169
Matsui (1989), 98
Matsui (1991), 95, 120
Matsui (1992), 69
Matsui (2008), 145
Matsui (2017), 199
Matsui and Matsuyama (1995), 121
Matsui and Okuno-Fujiwara (2002), 107
Matsui and Shimizu (2007), 145, 147
Matsui et al. (2012), 6
Matsuyama (1991), 121
Matsuyama et al. (1993), 107, 109
Maynard Smith and Price (1973), 56, 57, 71,
    79, 82, 98
Merton (1949), 193
Ministry of Education Culture Sports Sci-
    ence and Technology (2015), 5
Ministry of Health, Labour and Welfare
    (2017), 5
Miyasawa (1961), 74
Monderer and Shapley (1996), 74
Myerson (1978), 69, 71

N
Nash (1951), 51, 55, 57, 69, 95, 160
Nassau (2018), 34

O
OECD (2003), 1
OECD (2013), 5
Okada (1996), 201
Okuno-Fujiwara (1988), 120
Oliver (1984), 11
Oliver (1996), 14
Oliver (2013), 11

P
Pearce (1984), 95
Peirce (1898/1992), 145, 147
Pepperberg (1994), 145
Plato (1941), 162

R
Rabin (1990), 105
Robson (1990), 120

S
Sacks (1996), 35
Sandholm (2010), 52, 60
Schilpp (1979), 156
Selten (1975), 69, 71
Shapley (1962), 74, 77
Shapley (1964), 59
Shimono (2000), 31
Simon (1957), 145
Smith (1988), 133
Smith (2010), 65
Steiner (2016), 19
Sugino (2007), 14
Swinkels (1992), 81, 84

T
Taylor and Jonker (1978), 59, 79
Thomas (1985), 57
Thorndike (2017), 145

U
United Nations (2019), 124

V
Van Damme (1987), 52, 59
Van Damme (1991), 79, 80
Van Damme (1994), 52
Vega-Redondo (1996), 52
Von   Neumann   and   Morgenstern
    (1947/2007), 10, 54

W
Weibull (1997), 52

Y
Yamamori et al. (2008), 106
Yamamori et al. (2010), 106

Yoshino (2016), 25, 26
Young (1993), 91, 120

Z
Zeeman (1981), 59

Subject Index

A

Accessible(ility) (disability), 64, 66
Accessible(ility) (dynamics), 71, **71**, **74**, 75–
    77, 78, 79, 81, 86–88, 90, 97, 98, 100,
    100, 101, **101**, 103–105, 125, 126,
    134–137, 141, **175**, 175–177
    direct(ly), 97, 100, **100**, 101–105, **175**
    ε-, 71, 74, **74**, 78
Act
    Americans with Disabilities, 14, 66
    Basic — for Persons with Disabilities, 2
    for Eliminating Discrimination against
        Persons with Disabilities, 2
    on Employment Promotion etc. of Per-
        sons with Disabilities, 2, 3
    Services and Support for Persons with
        Disabilities, 23
Atelier Incurve, 41, 42, 45

B

Best response dynamics, 13, 69, 71, 74, 80,
    83, 86–88, 90, 114, 174, 201, 212
    socially stable set with respect to, SS
        set(BR), *see* socially stable set, with
        respect to best response dynamics (SS
        set(BR))
    socially stable strategy with respect to
        (SSS(BR)), *see* socially stable strategy,
        with respect to best response dynamics

C

Cheap-talk, **13**, 95–98, 104–106
    game with, *see* game, with cheap-talk
Chronic myelogenous leukemia (CML), 27,
    27, 28

Coherent(ce) (in inductive game theory),
    145, **148**, 151, 152, 157, 161, 162,
    179, 182–192, 195, 215, 216
    statistical, 149
    with active experiences, 183, 185, 196
    with experiences, 183
    with passive experiences, 183, 196
Colorblind (colourblind), 33–35
Complementarity, 65, 66
    institutional, 64, 65
    strategic, 51, 61, 124, 126, 200, 212, 213
Convention, 1–4, 13, 54, 61, 66, 107, **107**,
    108, 114, 115, 117–121, 123, 132
    eclectic, 13
    evolution of, 107, 113
    interaction of, 13, 107
    stable, 1
Convention on the Rights of Persons with
    Disabilities (CRPD), 2, 11, 21

D

Deaf, 6, 23–25, 199, 200, 202, 203, 210, 216,
    217, 219
    hereditary —ness, 14, 25, 199, 201, 202,
        212, 219
    non-, 25, 200, 203, 211, 212, 216
Disability, 14, 25, 34, 69, 159, 199, 215, 217,
    220
    adapted to, 201
    as a social construct, 11
    as endogenous institutions, 11
    association, 6, 63
    categories, 4
    children with and without, 216
    child with, 23, 24, 216
    disqualifying clause on, 35, 37

© Springer Nature Singapore Pte Ltd. 2019
A. Matsui, *Economy and Disability*, Economy and Social Inclusion,
https://doi.org/10.1007/978-981-13-7623-8

economy and —, *see* economy, and dis-
    ability
employment quota for persons with, 3
hearing, 7, 23, 24, 36
in Japan, 1, 4, 25
intellectual, 5, 7, 12, 37, 42–44
mental, 5, 7, 19, 44
non-, 14, 202
person with, 3–8, 10, 12, 14, 17, 19–21,
    24, 35, 36, 41–43, 45, 62–64, 66, 95,
    106, 200–202, 220
persons with and without, 20, 107, 215
physical, 5, 7
professional with, 37
registration system, 4
-related expenditures, 5
-related issues, 1, 2, 4, 10–14, 51, 52, 62,
    199, 217, 219
-related phenomena, 219
-related policies, 5
-related services, 5
self-studies of, 17
studies, 8, 11, 14, 201, 215
two streams of the concept of, 14
visual, 7
worker with, 106
Discrimination, 3, 4, 14, 35, 118, 119, 142,
    147, 159, 160, 169, 173, 190, 193
Disqualifying clause, 36, 37

E
Economics, 3, 8, 10–12, 14, 51, 124, 159,
    161, 179, 190, 193
behavioral, 146, 215
difference between natural science and,
    10
institutional, 8
mathematical, 10
neoclassical, 8, 194
Walrasian, 70
Economy
and disability, 1, 12, 13, 95, 215, 219, 221
exchange, 126
market, 18, 21, 65
planned, 21
scale, 61, 125
scale dis—, 61
Einstein, 147, 156, 157
Equilibrium, 13, 51, 52
analysis, 55
autarky, 126, 136
multiple, 53

Nash, *see* Nash equilibrium
no-coordination, 111, 112, 114
partial coordination, 113, 114
refinement, 56
segregation, *see* segregation, equilibrium
selection, 53
sequential, 190
subgame perfect, 153, 190, 206, 209, 216
unified, 114
Equilibrium evolutionarily stable set (EES
    set), 81
ES set(EE), *see* evolutionarily stable set
    (against equilibrium entrants) (ES
    set(EE))
Ethnicity, 160, 162–164, 169–171, 186–190,
    196, 197
configuration, 163, 164, 166, 167, 170,
    173, 181, 183, 185–187, 193
Evolutionarily stable set (against equilib-
    rium entrants) (ES set(EE)), **87**, 88
Evolutionarily stable set (ES set), 58
Evolutionarily stable strategy (ESS), 57, **57**,
    58, 59, 79, 80, 82, **82**, 83, 90
weak, 57, 58
Experience (in inductive game theory), 13,
    14, 145–148, **148**, 149–152, 155,
    157, 158, 160–163, 165, 166, **166**,
    167–169, 179, 181–186, 189–192,
    194, 195, 200, 201, 215, 216
active, 161, 162, 165, **166**, 167–169,
    182–185, 190, 192, 195, 196
passive, 161, 162, 165, **166, 167**, 168,
    182, 183, 185, 187, 192, 196
stationary, **166, 167**, 168, 182
Externality, 61, 65
negative, 126, 139
network, 64

F
Fictitious play, 59, 60, 71, 74, 77, 78
Shapley's example for, 78

G
Game
bargaining, 201, 203, 206, 208, 209
bargaining — with language constraints,
    200
coordination, 56, 71, 104, 120, 121, 125
dictator, 106
festival, 147, 159–161, 163–166, 169,
    174, 181, 183, 184, 190
form, 53

majority bargaining, 200, 203, 204, **206**, 207–210, 212, **212**, 214
majority bargaining — with language constraints, 203
of common interest with cheap-talk, 98, 101, 105
of matching pennies, 54, 77
prisoners' dilemma, 220
the battle of the sexes, 104
three-person bargaining, 203
two-person bargaining, 207
unanimity bargaining, 200, 203, 204, **210**, 211, 213, **213**, 214, 216
with cheap-talk, 97, 98, 101, 104, 175
without cheap-talk, 101
zero-sum, 54
Game theory, 1, 4, 8, 10–14, 51, 69, 70, 88, 89, 124, 146, 147, 154, 159, 160, 164, 179, 199, 215, 217, 219, 220
deductive, 160, 190, 194, 206, 215
evolutionary, 13, 51–53, 56, 57, 59, 69, 160, 212
inductive, 14, 146, 162, 190, 194, 199, 215, 217
noncooperative, 95, 214
three approaches in, 200
three approaches to, 161, 201

H
Hume, 147, 154, **154**, 155–157

I
Impairment, 11, 14, 25
hearing, 199, 216, 217
Independent living (life), 5, 12, 18, 20

J
juku, 12, 46, 47

M
Martha's Vineyard Island, 14, 25, 199, 201, 202, **202**, 211, 212, 217, 219
Model (in deductive game theory), 146
Model (in disability studies)
social, 11
social — of the United Kingdom, 14
Model (in inductive game theory), 14, 145–148, **148**, 149, 151–154, 157, 158, 161, 162, 167, 179, 181–185, 189–195, 200, 215, 216

construction of, 146–148, 150, 157
individual, 180, **180**, 181–185, 190, 191, 194
mere-enumeration, 162, 182, 191
naive hedonistic, 162, 179, 184, 185, **185**, 186, 187, 192, 193
of the world, 157
social, 194
sophisticated hedonistic, 162, 179, 184, **185**, 187, 189, 190, 192, 193, 196
true-game, 162, 181, **181**, 182–185, 190–193

N
Nash equilibrium, 13, 51, 55, **55**, 57, 58, 60, 69–71, **73**, 76–78, 81, 82, **82**, 84, 85, 95, 109, 110, **110**, 146, 149, 152, 163, 165, **165**, 168–170, 174, 190, 191, 214
interpretations of, 55, 69
refinements of, 52, 56, 71
strict, 90, 91, 110, 119
symmetric, 55, **55**, 57, 80
Nothing about us without us, 11, 21

O
Ordinary people, 11, 14, 23, 35

P
Pingelap, 35
Prejudice, 3, 12, 14, 33, 35, 66, 147, 159, 160, 162, 179, 184, 185, 187–190, 192–194, 199, 201, 214–217, 219
Price theory, 70

R
Rare disease, 26–31
Rational
agent, 51
behavior, 52, 53, 215
boundedly — agent, 53
player, 69, 160, 215
Rationality, 13, 51, 52, 69, 89
bounded, 215
hypothesis, 200, 214, 215
interpretation, 55, 56
sequential, *see* sequential rationality
Research on Economy And Disability (READ), 6, 33, 37
Research on Economy And Social Exclusion (REASE), 25

S
Sarcoma, 26
Segregation, 14, 147, 159, 169, 171, 172,
        174, 179, 188, 192, 219
    equilibrium, 171–174, 177, 185, 187–
        189
Sequential rationality, 169, 190–192, 196
Sign language, 23–25, 123, 199, 200, 202–
        205, 208, 209, 211–214, 216, 217,
        219
Socially stable set, 71, 73, 75, **75**, 76–79,
        81, 88–90, 97, 98, 101, **101**, 104, 105,
        163, 174, 175, 177, 214
    with respect to best response dynamics
        (SS set(BR)), **87**, 88, 175
Socially stable strategy, 71, 73, **75**, 76, 77,
        81, 86, 90, 101

against equilibrium entrants (SSS(EE)),
        81, **84**
    with respect to best response dynamics
        (SSS(BR)), 80, **84**
Social norm, 52, 53
    as an equilibrium, 52
SSS(BR), *see* socially stable strategy, with
        respect to best response dynamics
        (SSS(BR))

V
Voice, **13**, 62, 63, 95, 106

Z
Zorn's lemma, 76, 77

Printed in the United States
By Bookmasters